Water in Wood

SYRACUSE WOOD SCIENCE SERIES, 4

SYRACUSE WOOD SCIENCE SERIES

Wilfred A. Côté, *editor*

1 John F. Siau, *Flow in Wood*, 1971.

2 B. A. Meylan and B. G. Butterfield, *Three Dimensional Structure of Wood*, 1972.

3 Benjamin A. Jayne, editor, *Theory and Design of Wood and Fiber Composite Materials*, 1972.

4 Christen Skaar, *Water in Wood*, 1972.

Water in Wood

CHRISTEN SKAAR

*State University College of Forestry
at Syracuse University*

SYRACUSE UNIVERSITY PRESS   1972

Library of Congress Cataloging in Publication Data

Skaar, Christen.
  Water in wood.

  (Syracuse wood science series, 4)
  Bibliography: p.
  1. Wood–Moisture. I. Title. II. Series.
TA419.S52   620.1'2   70-39754
ISBN 0-8156-5034-5

Foreword

It is appropriate that a fundamental book on wood-water relationships should appear at this point in the Syracuse Wood Science Series. Wood anatomy and ultrastructure are, of course, basic to all of the topics that will be included in this series. However, the behavior of wood resulting from changes in its moisture content would have to be considered of equal practical importance with knowledge of the structure of wood, the natural composite material. In fact, discussions of physical and mechanical properties of wood-based materials are incomplete (and invalid) if this critical question is overlooked.

Water in Wood, with its theme of wood and water interactions and their implications for the particular wood properties under discussion, provides a link between volumes in the Syracuse Wood Science Series. The author, Dr. Christen Skaar, has undertaken a most challenging task in this book—that of including all of the important physical phenomena relating to the behavior of wood in response to moisture-content changes and all of the related fundamental considerations necessary for a thorough understanding of these phenomena. He has managed this very well without sacrificing the conciseness that is so desirable. This success was not accidental as he has had more than two decades of experience through teaching and research in the physics of wood.

Syracuse, New York
Spring 1972

Wilfred A. Côté, *series editor*
Professor of Wood Technology
Director, Laboratory for Ultra-
    structure Studies
State University of New York
College of Forestry at Syracuse University

Preface

Wood is unlike most construction materials used by man in that it is produced biologically by a living organism, the tree. Therefore, in common with most biological materials, it is formed in an essentially water-saturated environment. After a tree is felled the wood begins to lose most of its moisture since it is no longer in continuous contact with water drawn up from the soil. Associated with this moisture loss are certain changes in the physical properties of the wood which are important with respect to its subsequent use. Such changes, and their relationships to the environment to which the wood is subsequently exposed, are the subject of this book.

The text material is divided into five chapters. The first chapter introduces certain necessary concepts concerning the physics of water. The second chapter treats the empirically observed relationships between wood moisture content and its environment. It also includes moisture-measuring methods with emphasis on electric moisture meters. The third chapter deals with the practical and theoretical aspects of the hygroscopic shrinking and swelling of wood, including anisotropic effects and the fiber-saturation point. The fourth chapter treats the thermodynamics of moisture sorption. The fifth chapter discusses some of the more important theories which have been proposed to explain the sorption isotherm for wood; that is, its equilibrium moisture content in relation to atmospheric conditions.

There has been no attempt to discuss wood structure or moisture movement in wood in detail since these are covered in the first book in this series, *Flow in Wood*, by John F. Siau (1970), and also by A. J. Stamm (1964) in *Wood and Cellulose Science*. The latter text also gives a thorough treatment of the methods for dimensionally stabilizing wood against moisture changes, which is omitted in the present text. Also omitted is a detailed discussion of the theoretical aspects of the electrical properties of wood which will be covered in a later book in this series.

I am indebted to the State University College of Forestry at Syracuse University for providing the climate and facilities which made it possible for me to conceive of and complete this book. Thanks are due to my colleagues and students in the Wood Products Engineering Department for their assistance. The United States Forest Products Laboratory at Madison, Wisconsin, also provided facilities and services, particularly those of William T. Simpson who fitted the sorption data of that laboratory, as given in the *Wood Handbook* (1955), to several of the sorption theories given in Chapter 5. The Forestry Department of the National Taiwan University, Taipei, Taiwan, provided the opportunity for me to make progress on the manuscript while I was teaching there in 1970. During my stay in Venezuela, the Laboratorio Nacional de Productos Forestales, Merida, also contributed facilities and materials on which are based some of the results published here for the first time. Many of the line drawings have been adapted from various sources which are acknowledged on the individual figures. I would also like to acknowledge the interest of A. J. Stamm of North Carolina State University at Raleigh and of George Bramhall of the Western Forest Products Laboratory of Canada, Vancouver, both of whom provided helpful criticisms of the manuscript. Particular appreciation is given to Miss Judy Barton and Miss Barbara Ryan for typing the manuscript. Finally I wish to thank my wife Dorothy for her inspiration without which the manuscript could not have been completed.

Syracuse, New York   CHRISTEN SKAAR
Spring 1972

Contents

Water in Wood

1. The Physics of Water

In order to understand the interaction between water and wood it is necessary first to discuss some of the physical properties of water itself. We shall therefore consider some of these properties and derive certain equations by which they are interrelated. Some of these equations will also be useful in subsequent consideration of the wood-water system.

Humidity and Vapor Pressure of Water

Water can exist in three general states or phases—solid (ice), liquid, or vapor, depending upon the temperature and pressure to which it is exposed. In Figure 1.1 the phase diagram for water (not drawn to scale) is shown. It can exist in each phase only when the temperature and pressure are within the limits shown. At equilibrium two phases can exist only at the conditions of temperature and pressure represented by the line between the phases. At the triple point, all three phases may exist simultaneously. At the critical temperature ($374°C$) and pressure (218 atmospheres) there is no longer any distinction between liquid and vapor.

The extension of the liquid-vapor line below the freezing point actually represents the vapor pressure of supercooled liquid water. Notice that it is higher than the solid line representing the vapor pressure of ice. We will see later that this difference in vapor pressure between ice and supercooled water can be used to calculate the heat of fusion of ice, that is, the heat required to melt ice. The regions of primary interest to us in Figure 1.1 are those areas denoting the vapor state and the lines separating the vapor state from the liquid and, to a lesser extent, the solid state.

The primary difference between the liquid and vapor state is the spacing between the water molecules. In the liquid phase the molecules are sufficiently close to each other so that appreciable forces of attraction and repulsion exist among them. The individual molecules are constantly vibrating about a region of equilibrium where the force between adjacent molecules is zero. Figure 1.2 shows the relationship that may exist between two adjacent water molecules as a function of the spacing x between them. At the equilibrium distance x_0 the force is zero. This is about 3 Å (Angstrom units) for liquid water. At distances greater than x_0 the molecules are attracted to each other by attractive forces

1

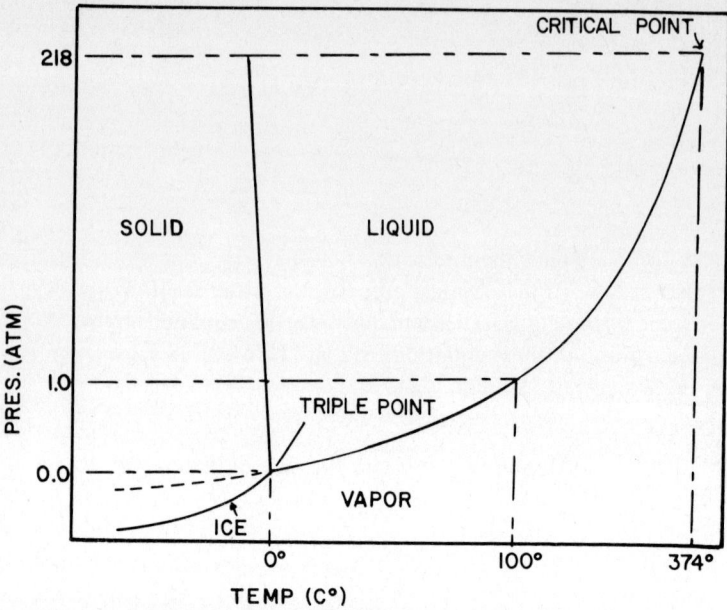

Fig. 1.1. Phase diagram for water (not to scale).

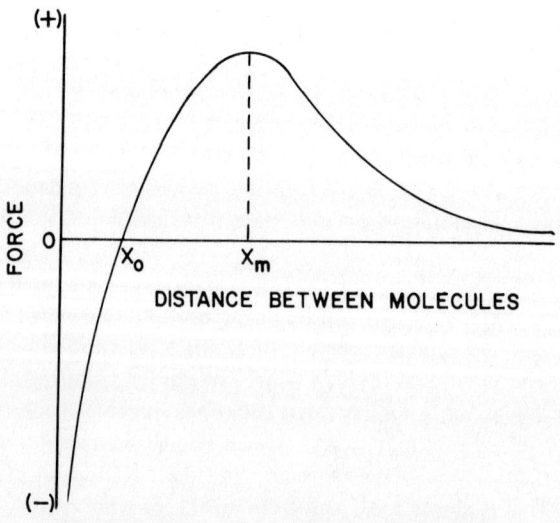

Fig. 1.2. Force-spacing relationships for water.

varying approximately as the inverse of the square of spacing x. At spacing less than x_0, strong repulsive forces operate which vary inversely as x^n, where n is a large number of the order of 4 to 10.

Most of the molecules in liquid water are vibrating to the left of x_m, the distance at which maximum attraction between molecules occurs. However, if a molecule attains enough energy to go beyond this point it will become a vapor molecule and escape from the force field of other water molecules. Only a small fraction of molecules have sufficient kinetic energy to break away from their neighbor molecules in the liquid form. These molecules become vapor molecules and, because of their high kinetic energy, exert a pressure against an enclosure which is called the vapor pressure of the water. The higher the temperature of the liquid water, the greater will be the number of water vapor molecules which have sufficient energy to escape from the liquid to the vapor state, and therefore the greater will be the vapor pressure of the water.

The mean or average distance between the centers of water molecules in liquid water can easily be calculated by use of Avogadro's number which states that there are 6.02×10^{23} molecules of any pure material per gram-molecular weight of the material. The gram-molecular weight of ordinary water is 18 grams per mole, and it occupies a volume of 18 cc at room temperature. The volume occupied by one water molecule in liquid form therefore is

$$\frac{\text{volume}}{\text{molecule}} = \left(\frac{18 \text{ cc}}{\text{mole}}\right)\left(\frac{1}{6.02 \times 10^{23} \text{ molecules/mole}}\right) = 30 \times 10^{-24}\left(\frac{\text{cc}}{\text{molecule}}\right).$$

The average distance between molecules therefore is approximately the cube root of 30×10^{-24} (assuming a cubic spacing), or slightly more than 3×10^{-8} cm, which is equal to 3 Å. This distance remains nearly constant with temperature since the density and specific volume of liquid water vary only to a small extent with temperature compared with the vapor state.

The average distance between water molecules in the vapor state can also be calculated similarly. However, it varies considerably with temperature and relative humidity because the density of water vapor varies with these factors. Table 1.1 gives the density of saturated water vapor at various temperatures as well

Table 1.1. Properties of Water Vapor

Temp. (°C)	Density of Saturated Water Vapor (g/cc)	Spacing Between Water-Vapor Molecules (cm)	Pressure of Saturated Vapor (p_0)	
			mm. Hg	Atm.
0	0.4846×10^{-5}	183×10^{-8}	4.579	0.00604
20	1.7291×10^{-5}	120×10^{-8}	17.535	0.0230
40	5.1164×10^{-5}	84×10^{-8}	55.324	0.0728
60	13.024×10^{-5}	61×10^{-8}	149.38	0.1965
80	29.338×10^{-5}	47×10^{-8}	355.1	0.469
100	59.77×10^{-5}	37×10^{-8}	760.0	1.000

as the mean distance between molecules and the saturated vapor pressure of the water.

It is clear from Table 1.1 that the spacing of water vapor molecules is from 12 to 60 times larger than the spacing of liquid water molecules, even when the air is saturated over the temperature range from 0° to 100°C. When the vapor pressure is lower than saturated at any temperature, the spacing is even farther apart.

It appears, to a first approximation, that a linear relation exists between the logarithm of the saturated vapor pressure of water ($\log p_0$) and the reciprocal, $1/T$, of absolute or Kelvin temperature, T, where $T = 273.1° + C°$ (Figure 1.3).

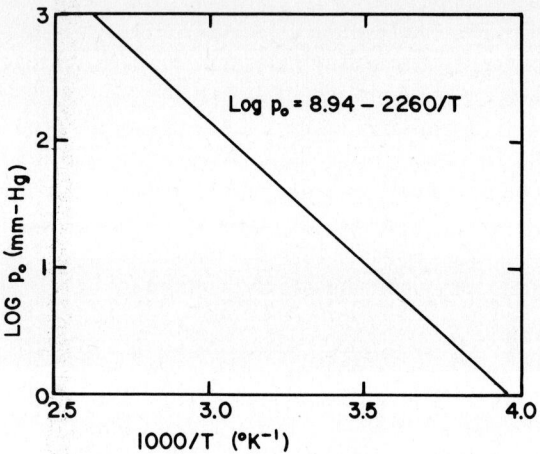

Fig. 1.3. Vapor pressure p_0 of water as a function of temperature.

From the data given in Table 1.1, the equation obtained is approximately

$$\log p_0 = 8.94 - (2260/T).* \tag{1.1}$$

It will be shown below that the slope of this curve is related to the heat required to evaporate a unit mass of water.

Up until now we have been discussing fully *saturated* air, that is, air which can hold no more water. Normally the ordinary atmosphere is not saturated, and the actual vapor pressure p is lower than the *saturated* vapor pressure p_0. The air humidity is often measured in terms of the relative vapor pressure h, which is defined as the ratio of the actual existing vapor pressure p to the saturation pressure p_0. Figure 1.4 shows the relationship between vapor pressure p, and tem-

---

*A more exact equation is the Kirchoff formula in the form

$$\log p_0 = A - B/T - C \log T$$

where A, B, and C are constants (Malmquist 1958).

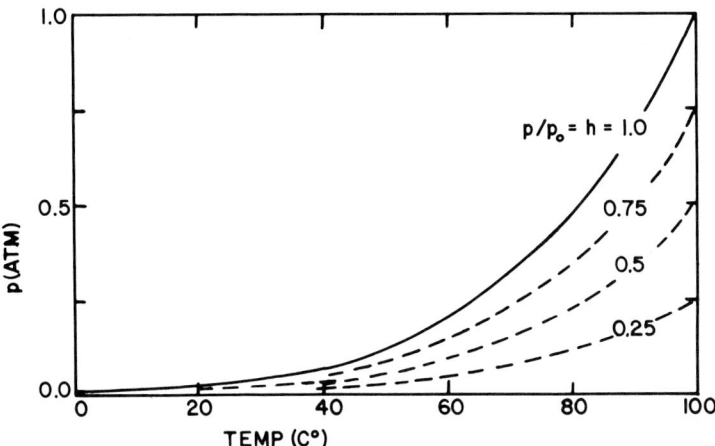

Fig. 1.4. Partial vapor pressure p as a function of temperature for different relative vapor pressures h.

perature for various relative vapor pressures. It is clear from Figure 1.4 that the relative vapor pressure h is dependent on the vapor pressure p and also the temperature, since this determines the value of p_0. In a room, for example, the vapor pressure p tends to remain constant throughout the room, but the temperature fluctuates considerably from one location to another. Therefore there may be wide variations in relative humidity within the room. For example, if the temperature variation within the room is one degree C, the relative humidity (relative vapor pressure multiplied by 100 percent) at room temperature ($25°C$) may vary by 3 percent within the room, even if the actual vapor pressure is constant within the room. It is therefore important to have good temperature control in a room which is designed to maintain constant humidity.

Under ordinary conditions, the relative vapor pressure h is less than unity, that is, the existing vapor pressure p is less than the saturated vapor pressure p_0. If the temperature of a room or of an object in the room is lowered, the saturated vapor pressure p_0 will also be lowered or reduced. There will be a temperature at which moisture will begin to condense from the atmosphere onto a cold object. The temperature at which condensation of moisture on an object begins is called the dewpoint temperature.

For example, referring to Figure 1.5, if the temperature in a dry kiln is $100°C$ and the relative vapor pressure h is 0.5, the dewpoint temperature is determined by the intersection of the horizontal broken line with the curve for $h = 1.0$. In the figure it appears to be about $82°C$. If the walls of the dry kiln are at a temperature of $82°C$ or lower there will be condensation of moisture on the walls, even when the relative humidity is as low as 50 percent.

Equation (1.1) can be used to calculate the dewpoint temperature from the

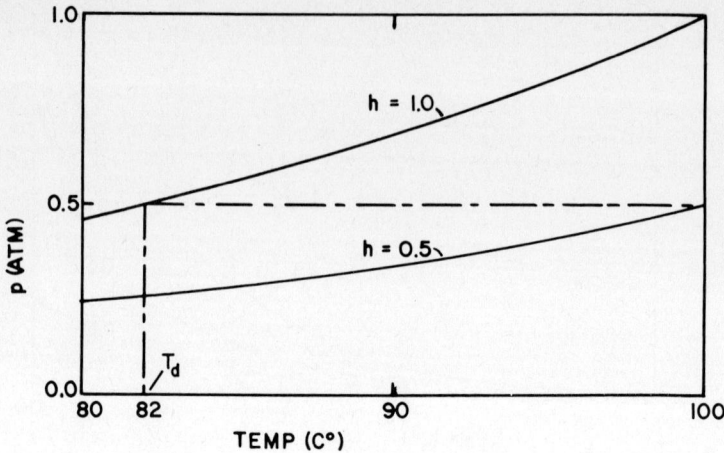

Fig. 1.5. Dewpoint temperature T_d in relation to vapor pressure p_0 and relative vapor pressure $h(p/p_0)$.

dry-bulb temperature and relative vapor pressure. This can be done by rearranging equation (1.1) into the form

$$p_0 = (10^{8.94})(10^{-(2260/T)}) \tag{1.2}$$

where p_0 is the saturated vapor pressure at the temperature T (Kelvin). Since the dewpoint temperature T_d is that temperature at which the vapor pressure p is the saturation vapor pressure $(p_0)_d$, we can also write

$$p = (p_0)_d = (10^{8.94})(10^{-2260/T_d}) \tag{1.3}$$

or combining (1.2) and (1.3)

$$p/p_0 = 10^{2260\ [(1/T)-(1/T_d)]} \tag{1.4}$$

or

$$\log(p/p_0) = 2.26\ [(1000/T) - (1000/T_d)] \tag{1.5}$$

or

$$\log(p_0/p) = 2.26\ [(1000/T_d) - (1000/T)]. \tag{1.6}$$

Equation (1.6) can be rearranged into a form which is most convenient for calculating the dewpoint temperature T_d if p/p_0 and the air temperature T are known. Thus, equation (1.6) can be written as

$$T_d = \frac{1,000}{(1000/T) + [\log(p_0/p)]/2.26}. \tag{1.7}$$

As an example of the use of equation (1.7), which is an *empirical* equation or experimentally obtained equation, let us work out the dewpoint temperature for the same case shown graphically in Figure 1.5, where T is $273.1 + 100° = 373.1°$ Kelvin, and p/p_0 is 0.5. Substituting these values into equation (1.7) results in $T_d = 355°K$, which, in degrees $C = 355 - 273.1 = 82°C$, the same answer as shown in Figure 1.5.

Similarly, if the dewpoint temperature, $T_d - 273.1$, and the air temperature, $T - 273.1$, are known, the relative vapor pressure p/p_0 can be calculated by use of equation (1.4).

It should be emphasized, however, that equations (1.1) to (1.7) are only approximate since there is some curvilinearity in the relationship between $\log p_0$ and $1/T$. For many practical calculations it is adequate.

Water Vapor and the Ideal Gas Law

At normal pressures and temperatures many gases, to a first approximation, obey the relationship called the ideal gas law. This relationship is:

$$pv = nRT = (w/M)\,RT \tag{1.8}$$

where p is the pressure of a gas, v is its volume, n is the number of moles of gas (the mass w, divided by molecular weight M), T is the absolute temperature in Kelvin degrees, and R is the universal gas constant whose value depends upon the units in which p and v are measured. The ideal gas law obeys the empirical observation called Boyle's law which was first formulated in 1660. Boyle's law states that the product pv of the pressure and volume of a gas is constant when temperature is constant. It also agrees with Gay-Lussac's (or Charles') law, observed in 1802, that the volume v of a gas varies linearly with absolute temperature T at constant pressure p.

The ideal gas law can also be derived from considerations of the kinetic theory under the assumption that the gas stores energy only in the form of translational kinetic energy, that there are no forces of attraction between gas molecules, and that the volume occupied by individual molecules is zero or negligible.

All real gases deviate from the ideal gas law, and water vapor is no exception. However, it can be shown that over the temperature range from $0°$ to $100°C$ the deviation is slight, in the order of 1 percent or less. At $0°C$, for example, saturated water vapor has a density ρ of 0.4846×10^{-5} g/cm^3, or 0.4846×10^{-2} g/liter, and a vapor pressure p_0 of 0.00604 atmosphere, from Table 1.1. Rearranging equation (1.8), and substituting in these values

$$R = (p_0 M)/(\rho T)$$

$$R = \frac{(0.00604)\,(18)}{(0.004846)\,(273.1)} = 0.0823 \text{ liter-atm-mole}^{-1}\text{deg}^{-1} \tag{1.9}$$

compared with 0.08205 for an ideal gas.

At 100°C (373.1° Kelvin) saturated water vapor has a density of 0.5977 g per liter, and the vapor pressure is one atmosphere. Therefore

$$R = \frac{(1.000)\,(18)}{(000.5977)\,(373.1)} = 0.0807 \text{ liter-atm-mole}^{-1}\text{deg}^{-1}$$

At 20°C, $R = 0.0819$; at 40°C, $R = 0.0818$; at 60°C, $R = 0.0816$; and at 80°C, $R = 0.0812$, in the same units.

Heats of Vaporization, Fusion, and Sublimation of Water

It has been shown earlier that the vapor pressure of water, p_0, changes rapidly with temperature. It will be shown here that this high rate of change is related to the energy required to evaporate water, the latent heat of vaporization. Likewise the vapor pressure of ice, and its variation with temperature, is related to the heat of sublimation of ice and to the heat of fusion of water into ice. The derivation of this relationship follows.

Consider that water can exist in three states and that each state is at a different potential energy level. A water molecule in the liquid state is in a potential energy "well," and in order to escape to a region of higher potential energy it must attain sufficient kinetic energy to overcome the attractive forces exerted by the nearby water molecules. This is analogous to a situation in which a man (Figure 1.6a) finds himself in a hole in the ground. If he is to escape from the hole he must give his body sufficient kinetic energy to jump out of the hole (Figure 1.6b) and thus overcome the potential barrier represented by the depth of the hole.

In the case of water molecules in the liquid form, only a small number of molecules at any instant have sufficient energy to escape from a water surface into the vapor state. This number increases as the water temperature is increased

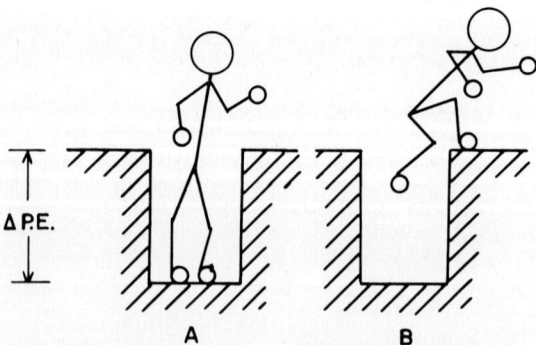

Δ P.E.

A   B

Fig. 1.6. Schematic diagram of a potential energy "well."

because of the increase in average kinetic energy associated with a temperature increase. The fraction N/N_T of water molecules which have sufficient kinetic energy to escape from a potential well whose depth is ΔPE can be expressed approximately by the Boltzmann distribution given as follows:

$$N/N_T = \exp(-\Delta PE/RT) \qquad (1.10)$$

where R is the gas constant and T the temperature (Kelvin).

Equation (1.10) can be shown graphically as in Figure 1.7, where N/N_T is

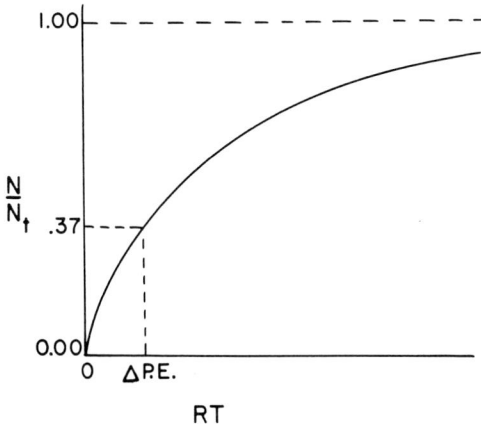

Fig. 1.7. Graphical representation of Boltzmann distribution.

plotted as a function of the product RT. Note that when the product RT is equal to the energy difference ΔPE between the two states, $1/e$ or about 37 percent of the molecules have sufficient energy to escape from the liquid to the vapor state.

It is more convenient in this case to express the gas constant R in heat units, or, $R = 1.987$ (approximately 2) calories per mole per degree Kelvin. The energy ΔPE which must be supplied to evaporate a mole of water is about 9,720 calories at 100°C (373°K). This is obtained by multiplying the latent heat of evaporation of liquid water (540 calories per deg C or K) at 100°C, by the molecular weight of water (18 grams per mole). The product RT at this temperature is only $(1.987) (373.1°) = 742$ calories per mole, so it is clear that RT is only a small fraction of ΔPE for water at ordinary temperatures. Therefore N/N_T is an extremely small fraction, in the order of only about 2 molecules per million even at the *boiling point* of water. This can be verified by substituting into equation (1.10) the appropriate values of RT and (ΔPE), as follows:

$$N/N_T = \exp(-\Delta PE/RT) = \exp(-9{,}720/742) = 2.05 \times 10^{-6}.$$

The vapor pressure p_0 of free liquid water at any given temperature is a measure of its escaping tendency at that temperature, which should be proportional to N/N_T. If this assumption is made, then

$$p_0 = K(N/N_T) \tag{1.11}$$

which, when substituted into equation (1.10), gives

$$p_0 = K\,[\exp(-\Delta PE/RT)]. \tag{1.12}$$

If we take the natural logarithm of both sides of equation (1.12) we obtain

$$\ln p_0 = \ln K - \Delta PE/RT. \tag{1.13}$$

Taking the derivative of both sides of equation (1.13) with respect to $1/T$ gives

$$\Delta PE = -2.303\,R\,[d(\log p_0)/d(1/T)] \tag{1.14}$$

assuming K is not affected by temperature and is therefore a constant.

Equation (1.14) can also be written in terms of calories of heat per *gram* of water evaporated, designated by the symbol Q_0, by dividing ΔPE by 18 g/mole, the molecular weight of water. Thus

$$Q_0 = \Delta PE/18 = -2.303(R/18)\,[d\log p_0/d(1/T)] \tag{1.15}$$

$$Q_0 = -0.254\,[d\log p_0/d(1/T)]. \tag{1.16}$$

Equations (1.14) and (1.16) are types of the Clausius-Clapeyron equation in differential form. They can be used to calculate the value of Q_0, the latent heat of vaporization of liquid water by taking the slope, $d(\log p_0)/d(1/T)$ of equation (1.1) and substituting into equation (1.15) or (1.16). Thus from equation (1.1), the slope $d(\log p_0)/d(1/T)$ is 2,260 degrees, and therefore from (1.16)

$$Q_0 = (-0.254)(2,260) = 574 \text{ cal/g water.}$$

Thus, it is possible, by means of the Clausius-Clapeyron equation, to calculate the heat of vaporization of free water if the variation of $\log p_0$ as a function of $1/T$ is known and if it is linear. The actual values of Q_0 obtained calorimetrically varies with temperature as is shown in Table 1.2. It is evident that the value of 574 cal/g water obtained above is in fair agreement with the mean value of 568 cal/g water obtained calorimetrically over the same temperature range.

When water freezes into ice its vapor pressure as a function of temperature also changes, as is shown by the discontinuity in the slope of the vapor pressure curve

Table 1.2. Values of Q_0 Obtained Calorimetrically

Temp. (°C)	0	20	40	60	80	100	Mean
Q_0 (cal/g water)	595.6	584.9	574.0	563.2	551.7	539.6	568

of Figure 1.1 at the triple point. Therefore the slope of log p_s against $1/T$, where p_s designates the saturated vapor pressure of solid water, or ice, is different from the slope of log p_0 against $1/T$. If one plots log p_s against $1/T$, a straight line results, the slope of which can be used to calculate the heat of sublimation of water from ice into the vapor state. Thus, equation (1.16) can be modified to find the heat of sublimation Q_s, or

$$Q_s = -0.254 \ \{(d \log p_s)/[d(1/T)]\}. \tag{1.17}$$

In general, Q_s is larger than Q_0, since $d(\log p_s)/d(1/T)$ is larger than $d(\log p_0)/d(1/T)$. The difference between Q_s and Q_0 is the heat of fusion Q_f, the energy required to melt one gram of ice. Thus

$$Q_f = Q_s - Q_0 \tag{1.18}$$

Substituting equations (1.16) and (1.17) into equation (1.18) results in

$$Q_f = -0.254 \left[\frac{d \log p_s}{d(1/T)} - \frac{d \log p_0}{d(1/T)} \right] \tag{1.19}$$

$$Q_f = -0.254 \left[\frac{d \log(p_s/p_0)}{d(1/T)} \right] \tag{1.20}$$

$$Q_f = +0.254 \left[\frac{d \log(p_0/p_s)}{d(1/T)} \right] \tag{1.21}$$

The data in Table 1.3 give the actual vapor pressures p_s and p_0 for ice and for

Table 1.3. Vapor Pressures of Ice and of Supercooled Water

Temperature (°C)	0°	-5°	-10°	-15°
Vapor Pressure of Water, p_0 (mm Hg)	4.579	3.163	2.149	1.436
Vapor Pressure of Ice, p_s (mm Hg)	4.579	3.013	1.950	1.241

supercooled water (water which is cooled slowly and carefully to prevent nucleation and resultant ice formation). Calculation of the heat of fusion Q_f from this data gives a value of 76 calories per gram of water, close to the value of 80 obtained calorimetrically.

The Capillary-Pressure Equation

The familiar capillary-pressure equation, $P_0 - P = 2\sigma/r$, can be derived by considering the energy or work required to expand the surface of an air-water inter-

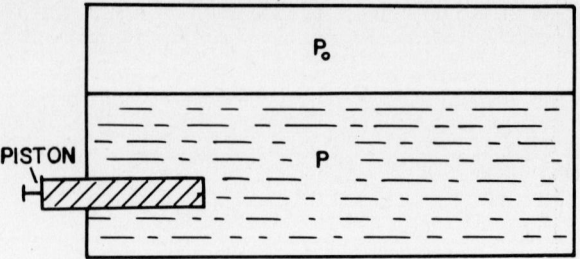

Fig. 1.8. Hydrostatic pressure P of water and of the atmosphere P_0.

face, such as an air bubble. Let P represent the total pressure in the water immediately under the surface, and let P_0 (not to be confused with p_0 which refers
to water vapor pressure) be the total pressure of the air above the water, including water vapor pressure. Under these conditions, represented by Figure 1.8,
$P = P_0$. Next place a permeable but rigid membrane with pores of radius r over
the surface of the water as shown in Figure 1.9. If the pressure P is reduced by

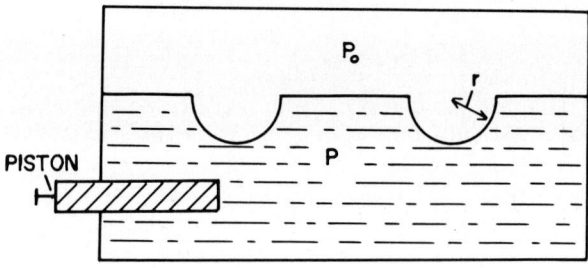

Fig. 1.9. Reduced hydrostatic pressure P of water compared with P_0 when radius r of the
air-water interface is less than infinity.

withdrawing the piston slightly, the water surfaces in the pores recede into the
water and, at a certain value of P, become spherical. Work must be performed as
the water surface or meniscus is drawn into the pores because of the surface
tension σ of the water, which resists increase of the area of the air-water meniscus
or interface.

Surface tension σ is defined as the work W required to increase the area A of
the surface, or $\sigma = dW/dA$. The increment of work dW required to increase the
area A by the increment dA is

$$dW = \sigma \, dA. \tag{1.22}$$

When the surface becomes a hemisphere, then $A = 2\pi r^2$, and $dA = 4\pi r \, dr$, and
equation (1.22) becomes

$$dW = 4\pi \sigma r \, dr. \tag{1.23}$$

But this increment of work dW is also equal to the work done by the difference in pressure $(P_0 - P)$ over the distance dr and area A, since work equals force times distance. The force is $(P_0 - P)A$, and the distance is dr; therefore,

$$dW = (P_0 - P) A \, dr = (P_0 - P) (2\pi r^2 \, dr). \tag{1.24}$$

Equating equations (1.23) and (1.24) results in the capillary pressure equation

$$P_0 - P = 2\sigma/r. \tag{1.25}$$

It should be noted that the water pressure P is always less than the total air pressure P_0, and the water really is in tension compared with the water shown in Figure 1.8. Therefore the term $P_0 - P$ is sometimes called the *capillary tension* which is more descriptive than the term *capillary pressure* from the point of view of the condition of the water.

The surface tension σ for the air-water interface decreases with increasing temperature, as shown in Table 1.4. The surface tension of the undiluted sap in

Table 1.4. Surface Tension of the Air-Water Interface

Temperature (°C)	0°	25°	50°	75°	100°	374° (Critical Temp.)
Surface Tension $\sigma \left(\dfrac{\text{dynes}}{\text{cm}} \right)$	75.6	71.8	67.9	63.5	58.8	0

green wood of four species measured by Stamm and Arganbright (1970) ranged from 45.5 to 57.3 dynes/cm at a temperature of 23 ±1°C, considerably lower than the value of 72 dynes/cm for distilled water at the same temperature.

Vapor-Pressure Depression by Capillaries (Kelvin Equation)

The tension in the water immediately below the surface of an air-water meniscus causes a reduction in the vapor pressure of the water, which can be related to the radius r of the meniscus. The equation which quantitatively describes this reduction in vapor pressure is known as the Kelvin or Thompson equation or simply as the vapor-pressure depression equation.

In order to derive this equation we use considerations similar to those used in deriving the capillary pressure equation, resulting in equation (1.23). For example, referring to Figure 1.10, if we have a permeable membrane with pores of radius r over a water surface in a closed container, the vapor pressure p above the water surface will approach p_0, the saturated vapor pressure, and the surface of the air-water meniscus will be flat, as shown by the broken line across the pore. If we now lower the vapor pressure of the water in the upper space by opening

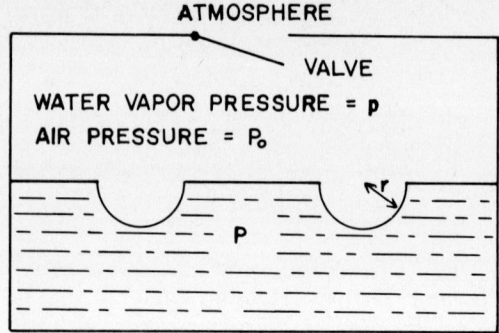

Fig. 1.10. Development of curved air-water meniscus of radius *r* resulting from reduction of partial vapor pressure *p*.

the valve (as shown in Figure 1.10) to an outer atmosphere which has a lower vapor pressure than p_0, water will evaporate through the pores, and the flat meniscus will curve inward to form a curved meniscus as shown by the solid curved lines in Figure 1.10. This results in increasing the surface area A of the water surface in much the same way as withdrawing the piston in the closed system shown in Figure 1.9. Work is done against surface tension in this case also, where $dW = \sigma\,dA$, and at the point where the surface becomes equal to that of a hemisphere, the increment of work dW required to increase the radius r of the hemisphere by the increment dr is similar to that given by equation (1.23), or $dW = 4\pi\sigma\,r\,dr$.

The work done in expanding the surface by the increment dA in this case is equal to the work done by the water vapor in expanding from the volume v_0, which it occupies at the saturated pressure p_0, to its new volume v, which it occupies at the reduced vapor pressure p. This work can be calculated if one assumes that water vapor obeys the ideal gas law, $pv = nRT$. For example, if w grams of water vapor expands from the volume v_0 at saturated vapor pressure p_0 to the volume v at pressure p, then the work W done by the vapor on its surroundings is given by

$$W = \int_{v_0}^{v} p\,dv = \int_{v_0}^{v} (w/18)RT(dv/v) \qquad (1.26)$$

since $p = (wRT/18v)$, where 18 is the molecular weight of water, and $n = w/18$, provided the expansion is isothermal at the temperature T. This work W done by the vapor is equal to the decrease in free energy ΔG of the water vapor, since when work is done by a gas its free energy or ability to do work is decreased by the amount of work which is extracted from it or which it does on its surroundings.

Integrating equation (1.26) results in

$$W = (w/18)RT(\ln v - \ln v_0) = (w/18)RT \ln(v/v_0) \qquad (1.27)$$

or since $pv = p_0 v_0$, and $v/v_0 = p_0/p$ at constant T, equation (1.27) becomes

$$W = w \, \Delta G = (w/18)RT\ln(p_0/p) \qquad (1.28)$$

where $w \, \Delta G$ is the decrease in Gibbs's free energy or the work done when a mass of w grams of water vapor expands isothermally from its volume at saturated vapor pressure p_0 to its volume at the lower vapor pressure p. Equation (1.28) or some variation of it will be used later in this text in conjunction with the thermodynamic properties of moisture held in wood.

Returning to our original objective of deriving the Kelvin or vapor-pressure depression equation for capillaries, we can modify equation (1.28) to calculate the work done, dW, when a small increment of dw grams of water expands from vapor pressure p_0 to p. Equation (1.28) then becomes

$$dW = (RT/18)[\ln(p_0/p)] \, dw. \qquad (1.29)$$

The incremental mass dw of water which expands is now taken to be equivalent to the incremental mass of water which evaporates when the radius r of the hemisphere increases by the small increment dr. Thus dw is equal to the increment dV of the volume of the hemisphere, multiplied by the density ρ of the liquid water, or $dw = \rho \, dV$, and equation (1.29) becomes

$$dW = (\rho RT/18)[\ln(p_0/p)] \, dV. \qquad (1.30)$$

However, the volume of a hemisphere $V = (2/3)\pi r^3$, and $dV = 2\pi r^2 \, dr$, so equation (1.30) becomes

$$dW = 2\pi \, (\rho RT/18)[\ln(p_0/p)]r^2 \, dr. \qquad (1.31)$$

This increment of work dW is equal to the work increment dW given by equation (1.33), which is the work done in expanding the meniscus by the increment dr. Thus, equating (1.23) and (1.31), we obtain the result

$$\ln(p_0/p) = [2\sigma(18)]/(\rho R T r) \qquad (1.32a)$$

which is the Kelvin equation for the vapor-pressure depression by a capillary of radius r for an air-water interface. This equation may be written in the alternate forms

$$p_0/p = \exp[(2\sigma/r)(18/\rho RT)] \qquad (1.32b)$$

or

$$p/p_0 = \exp[-(2\sigma/r)(18/\rho RT)] \qquad (1.32c)$$

or

$$2\sigma/r = (\rho RT/18)[\ln(p_0/p)] \qquad (1.32d)$$

or

$$2\sigma/r = - (\rho RT/18) [\ln (p/p_0)] \qquad (1.32e)$$

where, in cgs (centimeter-gram-second) units

   r = radius of capillary meniscus (cm)
   σ = surface tension of air-water interface (dyne cm^{-1}, or ergs cm^{-2})
   18 = molecular weight of water (g $mole^{-1}$)
   R = gas constant (= 8.32×10^7 ergs $mole^{-1}$ deg^{-1})
   T = absolute temperature (°Kelvin)
   ρ = density of water (= 1.0 g cm^{-3})
p/p_0 = relative vapor pressure.

Equations (1.32 a, b, c, d, and e) relate the relative vapor pressure p/p_0 over a curved meniscus to the radius r of the meniscus. It is evident from these equations that the relative vapor pressure at equilibrium with a curved surface is less than 1.0. For example, if the surface is flat, as over a plain water surface, then $r = \infty$ (infinity) and the term on the right side of equation 1.32a becomes zero. Hence the term p_0/p on the left side must be unity, since $\ln(1) = 0$. If the surface is curved the term r is less than infinity and the term on the right must be greater than zero. Hence p_0/p must be greater than unity, and p/p_0 must be less than one.

It is possible to obtain an intuitive understanding of the Kelvin equation relating vapor pressure and radius of an air-water meniscus by considering the relative attractions of surface water molecules to neighboring liquid water molecules when the surface curvature changes. For example, when the air-water interface is flat there is a greater tendency for water molecules near the surface to escape, compared with the tendency to escape from a concave curved surface. Consider Figure 1.11 which shows the possible paths of escape of water molecules near the surface under the two conditions. When the surface is flat, molecules which leave the surface from almost any angle can escape, whereas when the surface is

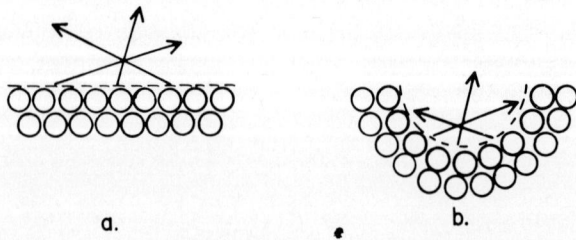

a.   b.

Fig. 1.11. Paths (arrows) of escaping water molecules near a flat surface a and a concave surface b.

concave curved those which leave from the surface at a low angle are likely to collide with the surface again and be recaptured. This means that fewer molecules will escape from a curved surface than from a flat surface, and the vapor pressure p over the curved surface will tend to be lower than the vapor pressure p_0 over a flat surface.

Another alternative viewpoint is that there are more neighboring surface molecules to a given surface molecule on the concave curved surface than on the flat surface, and that there is therefore a higher attractive force between adjoining molecules on the curved surface. The potential energy of a surface molecule is therefore at a lower level in a curved surface than in a flat one, and the kinetic energy which it must attain in order to evaporate is higher. This will also result in a lower vapor pressure.

The Kelvin equation also applies to convex surfaces such as raindrops or fog droplets. In this case, however, the relative vapor pressure p/p_0 is greater than unity, since the escaping tendency for molecules near the surface is greater for a convex curvature than for a flat surface. Equations (1.32) can be used to predict the vapor pressure elevation over a convex curved surface such as a fog droplet simply by using a negative value for the radius r of the droplet.

Swelling-Pressure (or Osmotic-Pressure) Equation

The capillary pressure equation (1.25) and the Kelvin equation (1.32) can be combined by eliminating the common term $2\sigma/r$ between them. This results in the swelling-pressure equation in the form

$$P_0 - P = (\rho RT/18) \ln(p_0/p) \tag{1.33}$$

which is also identical with the osmotic-pressure equation for solutions

$$\pi = (\rho RT/18) \ln(p_0/p) \tag{1.34}$$

where the osmotic pressure π of a solution is identical with $P_0 - P$, the capillary pressure.

Thus we see that capillary pressures and osmotic pressures are really similar phenomena in that they both relate the relative vapor pressure p/p_0 to the difference in hydrostatic or total liquid pressures between a solution and pure water in one case and between the water in a capillary and water in a large vessel in the other case. In the case of both aqueous solution and water in a capillary tube or under a curved air-water meniscus there is a reduction of vapor pressure and also a reduction of hydrostatic pressure. We will discuss these two phenomena briefly to show the relationships between them.

Consider the diagrams shown in Figure 1.12 in which a solution such as sugar in water is separated from pure water by a semipermeable membrane which permits water to pass through but which excludes sugar molecules. The value of

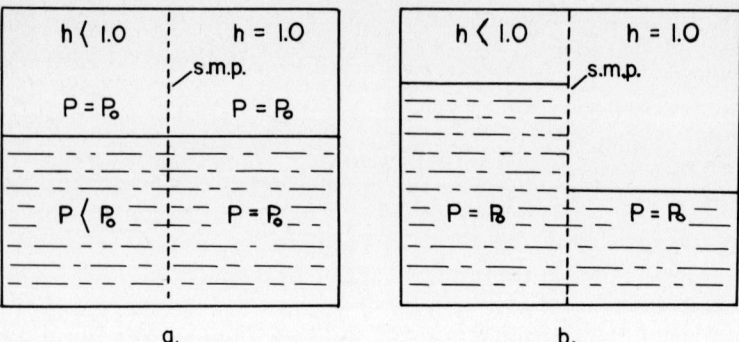

Fig. 1.12. Illustration of water flow into a solution separated by a semipermeable membrane (smp): a. initial condition; b. final condition.

p/p_0 above the sugar solution is less than 1.0, whereas that of the water is 1.0. If the total atmospheric pressure P_0 above both liquids is constant, the total hydrostatic pressure P in the solution is lower than P_0, the pressure in the liquid water. Therefore there is a pressure differential $P_0 - P$ which tends to draw water through the semipermeable membrane. This process continues until the level of the solution is sufficiently high above that of the water so that the weight of solution increases its pressure at the water level such that it is equal to the water pressure, and the system is at equilibrium. There is also a simultaneous flow of water vapor from the atmosphere above the water into the atmosphere above the solution because the vapor pressure p_0 of the water is greater than the vapor pressure p of the solution. If the membrane between the two liquids is impermeable there will be the same process of vapor movement so long as the atmospheres are open to each other, and the end result will be the same as in Figure 1.12b.

The same kind of process takes place when two fine capillaries of different size are subjected to the same atmospheric pressure P_0, as is shown in Figure 1.13. Suppose the two capillaries are originally isolated as in Figure 1.13a, and that the water level in each is originally the same. If capillary (1) is larger than capillary (2), its vapor pressure p_1 is higher than the vapor pressure p_2 of capillary (2). Hence water vapor will tend to evaporate from capillary (1) and diffuse through the air and recondense in capillary (2), thus raising the level of (2) and lowering that of (1). This process continues until the level in capillary (2) reaches the top of the tube. When this happens, the radius of meniscus (2) increases, approaching that of (1) until an equilibrium is attained at an intermediate radius. At this point the vapor pressure p_2 is slightly lower than p_1 because of the small effect of elevation on the vapor pressure. Furthermore, at equilibrium the hydrostatic pressure P_1 and P_2 are the same at any given level in the two tubes, as was the case in the osmotic solution situation shown in Figure 1.12b.

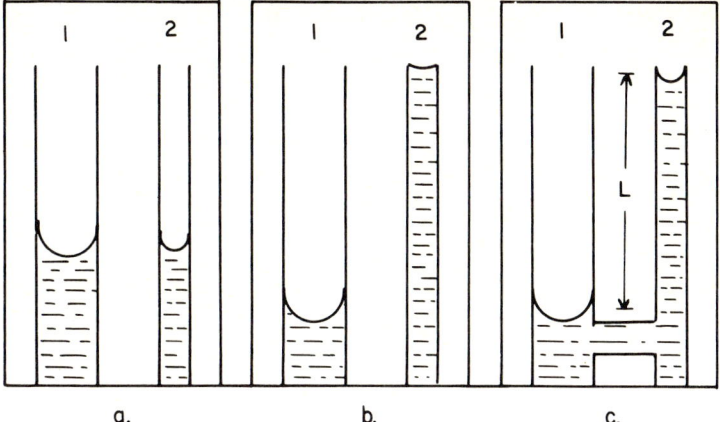

Fig. 1.13. Migration of water from a large to a small capillary tube: a. initial conditions; b. final condition for vapor movement; c. final condition for liquid movement where capillaries are connected.

If the two capillaries are interconnected by a tube, as shown in Figure 1.13c, a similar equilibrium condition is approached. In this case liquid water flows from capillary (2) into capillary (1) because the hydrostatic pressure P at the level of the connecting tube is higher in (2) than in (1). The same equilibrium condition is obtained as in the previous case, but more rapidly because of the higher flow rate of liquid water compared with vapor.

If the two capillary radii are r_1 and r_2, then the maximum possible difference L in the heights of the menisci in the two tubes is given by

$$L = (2\sigma/\rho g) \left[(1/r_2) - (1/r_1)\right] \cos\theta \qquad (1.35)$$

where g is the acceleration due to gravity, and θ is the wetting angle between the water and the capillary tube. In wood, water normally forms an angle θ of $0°$ unless there are some hydrophobic extractives in the wood, and therefore $\cos\theta = 1$. Likewise, the difference in hydrostatic pressures $P_1 - P_2$ in the two tubes is given by

$$P_1 - P_2 = \rho g L = 2\sigma \left[(1/r_2) - (1/r_1)\right] \cos\theta. \qquad (1.36)$$

When $\cos\theta = 0$, equation (1.36) reduces to the simpler form

$$P_1 - P_2 = \rho g L = 2\sigma \left[(1/r_2) - (1/r_1)\right]. \qquad (1.37)$$

The magnitude of capillary pressure as a function of capillary radius r can be calculated from equation (1.25) or some variation of it such as (1.37). Likewise, the magnitude of the relative vapor pressure (p/p_0) as a function of r can be computed from equation (1.32), or swelling pressure can be related to (p/p_0) by equation (1.34) or (1.33). The interrelationships of these three factors are

Table 1.5. Interrelationships of p/p_0, $P_0 - P$, and Capillary Radius r

$\dfrac{p}{p_0}$	Capillary Pressure ($P_0 - P$), or Capillary Tension, or Osmotic Pressure, etc.			Capillary Radius (of meniscus) Between Water and Air	
	Atmospheres	Dynes/cm^2	Cm Hg	Centimeters	Ångstroms
1.000	0.0	0.0	0.0		
0.9999	0.13	0.132×10^6	9.9	10.9×10^{-4}	109,000
0.999	1.3	1.32×10^6	99.0	1.09×10^{-4}	10,900
0.99	14	14.2×10^6	1060	0.1050×10^{-4}	1,050
0.9	142	144×10^6	10,800	0.0101×10^{-4}	101(?)*
0.8	400	406×10^6	32,000	0.0036×10^{-4}	36(?)
0.7	477	484×10^6	36,300	0.0030×10^{-4}	30(?)
0.6	680	690×10^6	51,700	0.0021×10^{-4}	21(?)
0.5	917	930×10^6	69,800	0.0016×10^{-4}	16(?)
0.4	1201	1212×10^6	91,300	0.0012×10^{-4}	12(?)
0.3	1560	1580×10^6	119,000	0.0009×10^{-4}	9(?)

*It is questionable if the capillary radii calculated from the Kelvin equation (1.32) are valid for radii below about 100 Å, since the concept of surface tension loses its meaning when surface areas are only a few molecules larger than the sizes or the spacing of the water molecules on the surface, 3 Å. This is because surface properties are derived from classical statistical concepts which assume that large numbers of molecules are involved in surface energy measurements. Therefore one should use caution in applying the equations relating capillary pressure or relative vapor pressure to capillary radii for p/p_0 less than, say 0.90. However, this does not invalidate the relationships of swelling pressure to p/p_0 which can be used to much lower values, say to p/p_0 of 0.30.

summarized in Table 1.5 for $T = 300°K$ ($t = 27°C$), and for a wide range of (p/p_0) and radius r. As mentioned in the footnote to the table, there is question about the validity of using the vapor-depression equation to calculate radius r for p/p_0 or h much lower than 0.9 (r less than 100 Å) because the surface radii calculated then approach molecular dimensions (3 Å), and intermolecular rather than surface tension forces as conceived in capillaries are the governing factors to determine the vapor pressures.

The author believes that the concepts of surface tension and capillary pressures or tension can be applied on the level of cell cavity dimensions but questions its use in calculating the dimensions of capillaries within the cell wall except at (p/p_0) greater than about 0.9. Therefore we will restrict its use to those conditions where its validity is certain, such as in the discussion of capillary or "free" water from the cavity or cell lumen of one cell to the next in wood.

The equations derived above apply strictly to circular capillaries or openings. However, the openings within the porous structure of wood are usually not circular and may consist of a shape with more than one radius. In this case equations (1.25) and (1.32) can be modified to give the effective radius r in terms of the two principal radii of curvatures, say r_1 and r_2. In this case the term $2\sigma/r$

is replaced by $\sigma((1/r_1) + (1/r_2))$, and equations (1.25) and (1.32d), for example, become

$$P_0 - P = \sigma [(1/r_1) + (1/r_2)] \qquad (1.38)$$

and

$$\sigma((1/r_1) + (1/r_2)) = (\rho RT/18) \ln (p_0/p). \qquad (1.39)$$

When $r_1 = r_2 = r$, these reduce to the original equations (1.25) and (1.32d).

At large values of p/p_0 it can be shown that a nearly linear relationship exists between p/p_0 and $P_0 - P$, or $1/r$, the reciprocal of capillary radius. In order to show this simplification, it is necessary to recall that the value of the exponential expression e^x can be represented by the following series expansion

$$e^x = \exp(x) = 1 + x + x^2/2! + x^3/3! + \cdots + x^n/n! \qquad (1.40)$$

This series is rapidly convergent for small values of x, and if x is sufficiently small, the equation reduces to

$$\exp(x) \approx 1 + x. \qquad (1.41)$$

For example, equation (1.32c) reduces to

$$h = p/p_0 \approx 1 - (36/\rho RT) (\sigma/r) \qquad (1.42)$$

is the approximation given by equation (1.41). It can be shown that if $x = -0.1$, then $\exp(x) \approx 1.0 - 0.1 \approx 0.9$, within half of one percent of the exact value of $\exp(x) = 0.90484$. For values of x closer to zero the approximation is even better. Therefore, the approximation given by equation (1.42) is valid for values of p/p_0 or h above 0.9 within the limitations specified above.

In terms of the radius r it can be seen from Table 1.5 that values of r which are larger than 100 Å, or 0.01×10^{-4} cm, or 0.01 micrometers (also called *microns* in older literature), give values of p/p_0 of 0.9 or larger. Therefore equation (1.42) can be used for calculations in this region of radii, essentially the same as the range of radii over which the Kelvin equation itself can be presumed to be valid.

It is convenient to convert equation (1.42) into the form

$$1 - h \approx (36/\rho RT) (\sigma/r) \qquad (1.43)$$

and to plot this relationship graphically as is shown in Figure 1.14, where curves are presented of $H(=100h)$ against $1/r$. Also shown are similar curves relating H and $(P_0 - P)$. In these curves the radius r is given in terms of micrometers since this is a convenient unit to use for the capillary structure of wood at the cell lumen or cavity level.

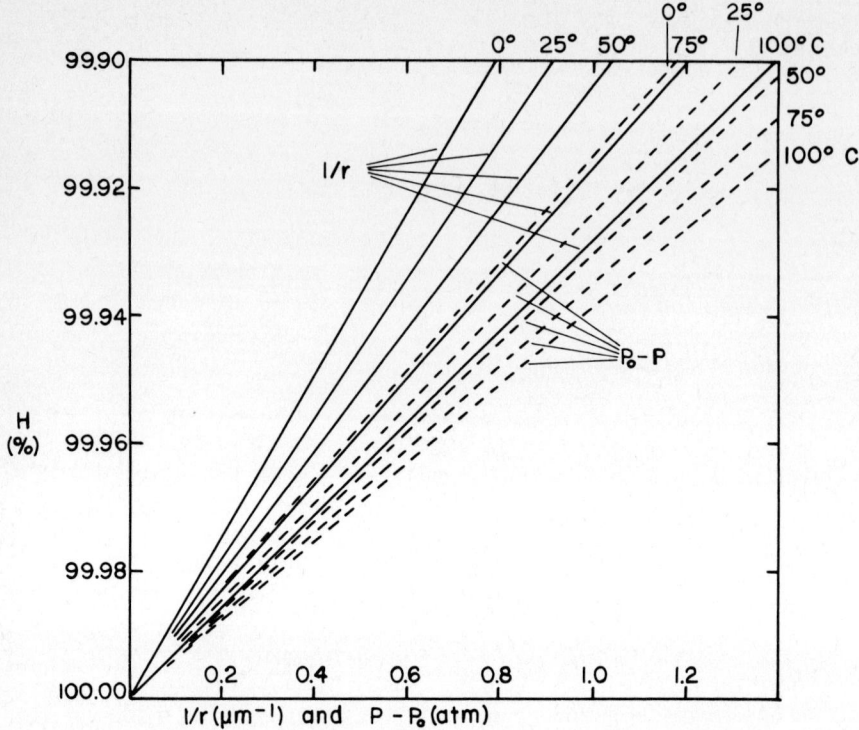

Fig. 1.14. Linear curves of percent relative humidity H against the reciprocal of capillary radius r (μm) and against capillary pressure $P_0 - P$ (atm) for temperatures from 0° to 100°C, in the humidity range above 99.9 percent.

Heat of Sorption of Capillary Water

We have already shown that the vapor pressure p over a capillary surface is lower than that of saturated water and also that this lowering of the vapor pressure compared with a free water surface is an inverse function of the capillary radius r. It is the purpose of this portion of the text to derive an equation which can be used to calculate the heat energy which is given up when water condenses into small capillaries or when it is taken up in liquid form from an open water surface by small capillaries.

In order to obtain this relationship, it is convenient to begin with the Clausius-Clapeyron equation (1.15) in the form

$$Q_c = (R'/18) \{d \, [\ln(p_0/p)] \, /d \, (1/T)\} \tag{1.44}$$

where Q_c is the *heat of sorption* in calories per gram of water taken up in liquid form by the capillary from liquid water, R' is the gas constant in *calories* mole^{-1}

degree^{-1}, and (p/p_0) and T are the relative vapor pressure h and temperature ($^\circ$K) of the capillary water. The rate of change of $\log(p_0/p)$ with $1/T$ for capillary water can be obtained from the Kelvin equation in the form given by equation (1.32d) written in the following manner

$$\ln(p_0/p) = 18/(\rho R''T)\,(2\sigma/r) \tag{1.45}$$

where R'' is the gas constant in dyne-cm mole^{-1} deg^{-1}. The derivative of $\ln(p_0/p)$ with respect to $(1/T)$ for a constant capillary radius r (equivalent to constant moisture content in a capillary), and assuming the density ρ of water to be constant, is

$$\frac{d\ln(p_0/p)}{d(1/T)} = \frac{36}{\rho R''r}\frac{d(\sigma T)}{d(1/T)}$$
$$= \frac{36}{\rho R''r}(\sigma - T(d\sigma/dT)). \tag{1.46}$$

Substitution of equation (1.46) into (1.44) gives

$$Q_c = \left(\frac{R'}{R''}\right)\left(\frac{2}{\rho r}\right)(\sigma - T(d\sigma/dT))$$
$$= \left(\frac{0.478 \times 10^{-7}}{r}\right)[\sigma - T(d\sigma/dT)]\ (\text{cal/g water}). \tag{1.47}$$

If we assume that σ varies linearly with temperature,* and that $d\sigma/dT = 0.168$ dyne cm^{-1} deg^{-1}, then, over the range from 0° to 100°C the term $(\sigma - Td\sigma/dT)$ is constant with a value of approximately 122 dyne/cm. Substituting this value into equation (1.47) reduces it to

$$Q_c \approx (0.478 \times 10^{-7})\,(122/r) \approx (5.83 \times 10^{-6})/r\,(\text{cal/g water}). \tag{1.48}$$

Table 1.6 gives the heats of sorption for capillaries of different radii r as calculated by equation (1.48) over the temperature range from 0° to 100°C.

Table 1.6. Heats of Sorption of Water in Capillaries of Different Radii

	Capillary Radius r				
Micrometers (microns)	100	10	1.0	0.1	0.01
Centimeters	10^{-2}	10^{-3}	10^{-4}	10^{-5}	10^{-6}
Heat of Sorption					
Q_c(cal/g)	0.000583	0.00583	0.0583	0.583	5.83

*The surface tension $\sigma \approx 75.6 - 0.168t$ from the data in Table 1.4, where t is in degrees C. Therefore $(d\sigma/dT) \approx 0.168$ dyne/cm-deg. Strictly speaking σ does not vary linearly with temperature over the range 0° to 100°C. A closer estimate of $(\sigma - T(d\sigma/dT))$ is 117, 118, 122, 126, and 133 at temperatures of 0, 25, 50, 75 and 100°C, respectively.

It is clear from Table 1.6 that the heat given up when wood is wetted above fiber-saturation, where most of the capillary water is held, is very small compared with the heat associated with bound-water sorption Q_L. This is particularly true of the gross capillaries formed by the cell cavity system, since these range from about one micrometer and upwards. However, the table does indicate that there is a finite heat of sorption even in the cell cavities.

The free-energy change ΔG_c and the entropy change $T\Delta S$ can also be calculated for capillary water. The free-energy change per gram of water is obtained from equation (1.22)

$$\Delta G_c = (R'T/18)\,[\ln(p_0/p)] \tag{1.49}$$

which, combined with equation (1.45), gives

$$\Delta G_c = (R'/R'')\,(2\sigma)/(\rho r). \tag{1.50}$$

Likewise, the entropy change associated with capillary water condensation from *liquid* water is equal to

$$T\Delta S = Q_c - \Delta G_c = (R/R'')\,(2/\rho r)\,[-T(d\sigma/dT)]. \tag{1.51}$$

At a mean temperature of 50°C it is possible to tabulate values of ΔG_c and of $T\Delta S$, if we take $\sigma = 67.9$ dynes cm^{-1} and $T(d\sigma/dT) = -54.1$ dynes cm^{-1}, since equations (1.50) and (1.51) reduce to

$$\Delta G_c = (0.478 \times 10^{-7})\,(\sigma/r)\ \text{cal/g water} \tag{1.52}$$

and

$$T\Delta S = (0.478 \times 10^{-7})\,(1/r)\,[-T(d\sigma/dT)]. \tag{1.53}$$

Table 1.7 lists the values of ΔG_c, $T\Delta S$, and Q_c at 50°C, taking the data given above at several different capillary radii as in Table 1.6. $T\Delta S$ accounts for 44.5 percent and ΔG_c for 55.5 percent of Q_c.

Table 1.7. Enthalpy, Entropy, and Free-Energy Changes of Water in Capillaries of Different Radii

at 50°C	Capillary Radius in Micrometers (microns)				
	100	10	1.0	0.1	0.01
Enthalpy, Q_c (cal/g)	0.000583	0.00583	0.0583	0.583	5.83
Entropy, $(T\Delta S)_c$ (cal/g)	0.000258	0.00258	0.0258	0.258	2.58
Free Energy, ΔG_c (cal/g)	0.000325	0.00325	0.0325	0.325	3.25

Free-Energy and Entropy Changes When Water Freezes

We have already calculated the heat of fusion Q_f for water freezing into ice, using equation (1.21), and have seen that it is in the order of 75 to 80 calories per gram of water. The free-energy changes ΔG_f and the entropy change $T \Delta S_f$ can also be calculated.

At the freezing point (0°C), the vapor pressure p_s of ice is the same as that of water and, from equation (1.48), $\Delta G_f = 0$. Therefore the entropy term $T \Delta S_f = Q_f$, and all of the energy removed from water when it freezes causes a decrease in entropy (an increase in order) of the water molecules. This also is true when water vapor at the vapor pressure p_0 condenses into liquid form and there is again a decrease in entropy and no free-energy change.

At temperatures below freezing, however, when supercooled water turns to ice at a given temperature there is a decrease ΔG in the free energy. This indicates that part of the total energy given up when supercooled water freezes results from a loss in free energy of the water. For example, at $-15°C$, p_0 and p_s are 1.436 and 1.241 mm of Hg (Table 1.3). Therefore $\Delta G = (1.987/18)(258.1)$ $\ln(1.436/1.241) = 2.1$ cal/g water, a small but finite decrease in free energy ΔG. This is the reason why super-cooled water is in an unstable or metastable state. A reaction or equilibrium always proceeds in a direction such that the free energy decreases.

2. Wood Moisture and the Environment

It is customary to express the moisture content of wood in terms of its dry weight using the equation

$$m = w_w/w_0 \qquad (2.1)$$

$$M = 100\,m \qquad (2.2)$$

where the fractional moisture content m is the ratio of the weight of water w_w to the dry weight of the wood w_0, and the percent moisture content M is the same ratio expressed as a percentage. Both these terms will be used throughout this text for convenience, depending on the context. The weight of water w_w is obtained by subtraction of the dry weight w_0 from the moist weight w_m of the wood, and equations (2.1) and (2.2) can therefore be written alternatively as

$$m = (w_m - w_0)/w_0 \qquad (2.3)$$

$$M = 100(w_m - w_0)/w_0. \qquad (2.4)$$

The moisture content of many materials is customarily given in terms of moist or wet weight rather than as the dry weight given above. The conversion between the two systems is given by the relationship

$$m_w = m/(1 + m) \qquad (2.5)$$

where m_w is the fractional moisture content based on moist weight.

Water in wood may exist in the cell wall as bound water or in the cell cavities in the form of vapor or liquid. The liquid water in the cell cavities is sometimes referred to as free water to distinguish it from the bound or hygroscopic water in the cell wall. It is not truly free water, however, since it is subject to capillary forces. Under ordinary conditions the weight of water vapor in the wood is negligible because of the low density of water vapor.

When wood dries, the water leaves the cell cavity of a particular cell first since the forces holding this water are appreciably lower than those holding the water in the cell wall. The moisture content at which the cell wall is fully saturated but the cell cavity is empty of liquid water has been designated as the fiber-saturation point, following Tiemann (1906). It is symbolized by $M_f(\%)$ or $m_f(\text{g/g})$. It usually ranges from 25 to 35 percent of the dry weight of the wood.

Its significance is discussed in Chapter 3. Stamm (1971) has critically examined nine methods for estimating the fiber-saturation point of wood.

Hydrogen Bonding and Water Sorption

Wood is a *hygroscopic* material; that is, it is able to remove water from the atmosphere and maintain a moisture equilibrium with the water vapor in the air. It is believed that wood is hygroscopic primarily because of the hydroxyl or OH groups which exist throughout its structure, particularly in the cellulosic and hemicellulosic portions of the wood. These hydroxyl groups attract and hold water molecules by the mechanism or chemical bond known as hydrogen bonding. Hydrogen bonding is also believed to be responsible for the partial crystallinity which is observed in the arrangement of the cellulose molecules in wood.

Hydrogen bonding occurs because the hydrogen atom is very small compared with the oxygen atom. Therefore, although the two atoms share an electron with each other, the small hydrogen nucleus with its positive charge is situated so close to the oxygen atom that it causes a residual positive charge on the near side of the oxygen atom. This results in a net negative charge on the opposite side of the oxygen atom which makes the atom electronegative on the opposite side from where the hydrogen atom is attached. For this reason the complex is polar; that is, it is not of a uniform charge throughout. The hydrogen side of the OH group has a positive charge and the oxygen side has a net electronegative charge. The positive side may therefore be considered to be a proton donor and the negative side a proton acceptor. This bonding potential is not strong compared with other primary chemical bonds, but it is nevertheless very important in materials which contain water.

The water molecule consists of two hydrogen atoms and one oxygen atom in the configuration shown in Figure 2.1. The angle between the two hydrogen atoms is about 105°. It is possible for a water molecule to form as many as four hydrogen bonds with other water molecules. In liquid water it is believed that

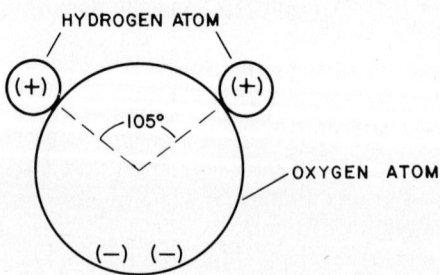

Fig. 2.1. Schematic diagram of a water molecule.

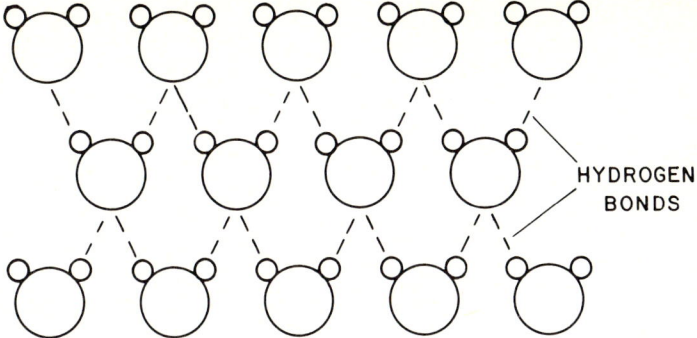

Fig. 2.2. Schematic diagram of the structure of ice showing hydrogen bonding (broken lines) between molecules.

groups of molecules are associated together into polymer-like aggregates which are loosely held together by hydrogen bonds.

In ice, on the other hand, a definite crystalline structure such as shown in Figure 2.2 is formed. The arrangement is tetrahedral and three-dimensional, with each molecule bonded to four other molecules by hydrogen bonds (indicated by the dotted lines). The distance between oxygen atoms is 2.76 Å, as shown in the figure. The strength of a hydrogen bond is approximately 5 kilocalories per mole.

Hydrogen bonding of water molecules to OH groups in wood and of OH groups to each other may occur in a manner similar to these bonds in ice or in liquid water. For example, we know that cellulose contains hydroxyl groups, as in Figure 2.3, which is a schematic diagram of a single repeat unit of cellulose (a glucose anhydride unit) in a cellulose chain. On this single unit are six OH

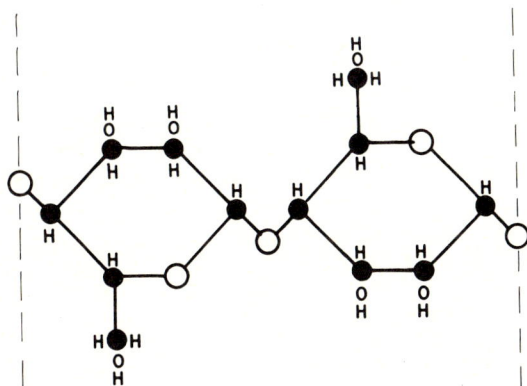

Fig. 2.3. Schematic diagram of a single repeat unit of the cellulose molecule.

groups available for hydrogen bonding. The upper half-unit is a mirror image of the lower half-unit. Therefore each half-unit contains three OH groups. The chemical formula of the cellulose molecule is usually given as $(C_6H_{10}O_5)_n$, where the basic unit is half of one of the repeat units shown in Figure 2.3.

A simple calculation can be made of the moisture content of cellulose when each of the OH groups is hydrogen-bonded to one molecule of water; that is, when the cellulose is fully hydrated. In this condition there are three water molecules per unit of $(C_6H_{10}O_5)$, giving the hydrated formula of $(C_6H_{10}O_5)$ · $3H_2O$. The moisture content of the hydrated cellulose based on its dry weight can then be calculated. One mole of dry cellulose weighs 162 grams, and three moles of water weighs $(3 \times 18 = 54)$ grams, giving a moisture content for fully hydrated cellulose of $(100 \text{ percent})(54/162) = 33$ percent.

This value of 33 percent is close to the value of the fiber-saturation point M_f for wood. However, for several reasons this is not to be interpreted as the reason why the cell wall can hold approximately this percentage of moisture. First, it is known that cellulose as found in wood is only partially accessible to water since much of it is in crystalline form which is inaccessible to water because the hygroscopic hydroxyl groups are mutually satisfied by the formation of hydrogen bonds between adjacent cellulose molecules in the crystalline regions (see Figures 2.4a and b). Second, there is evidence that more than one molecule of water is taken up or attracted to each accessible sorption site (hydroxyl group)— in fact, as many as five or six in the fully saturated cell wall. Finally, there are other important constituents of wood besides cellulose, including hemicelluloses, lignin, and various extractive materials which may have different sorptive properties than cellulose, as is discussed below.

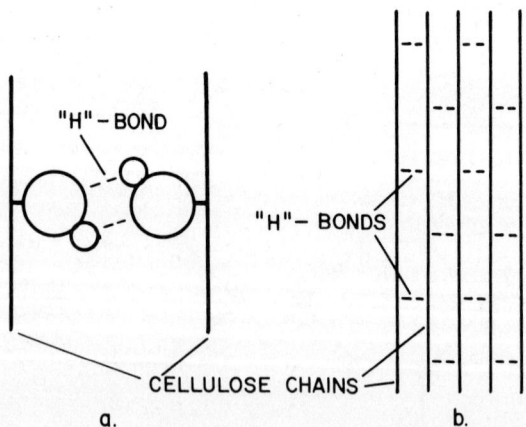

Fig. 2.4. Hydrogen bonds (broken lines) between cellulose molecules: a. single-bond pair between adjacent segments; b. series of hydrogen bonds linking several molecules in a crystalline region.

It is believed that in the crystalline regions of cellulose the chains form in a parallel, uniformly spaced arrangement, as shown in Figure 2.4b. The parallel chains are bound together at regularly spaced intervals by hydrogen bonds between hydroxyl groups. Individual cellulose molecules are longer than a single crystallite and may pass through more than one crystallite as well as through the unorganized noncrystalline or amorphous region, as shown in Figure 2.5. The

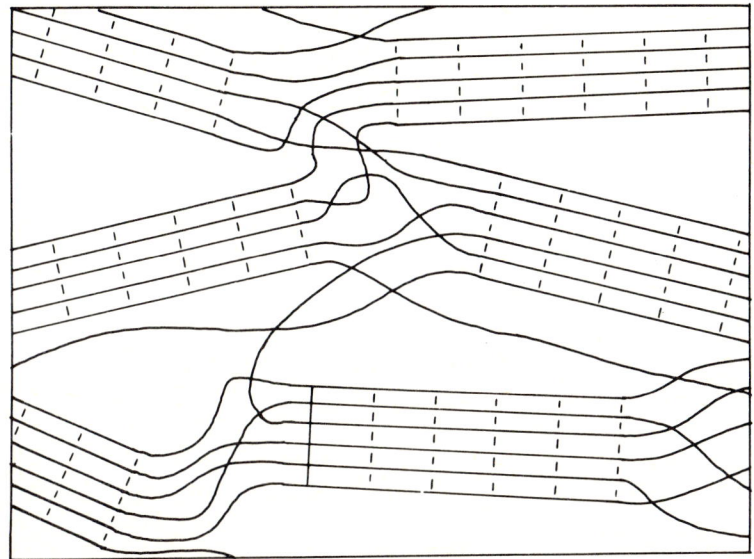

Fig. 2.5. Portion of cellulose showing ordered or crystalline regions and disordered regions.

OH groups in the crystalline regions are mutually bonded to adjacent OH groups from parallel chains and are therefore not available for bonding water. However, those on the surface of the crystallites or in the amorphous regions may be available for bonding water provided they are not cross-linked with other OH groups at points of contact.

According to evidence from the electron microscope the fine structure of cellulose is arranged into larger structures called fibrils and microfibrils. In some cases larger fibrils called macrofibrils are visible with the optical microscope.

Measurement of Wood Moisture Content

As the definition of wood moisture content implies, the basic method for measuring moisture content is gravimetric, but there are many—up to fifteen—

methods which have been used, according to Kollmann and Höckele (1962). All of these are not given here, and one or two additional methods are described which are not included among those mentioned by Kollmann and Höckele.

Gravimetric Method

In the basic gravimetric method of measuring wood moisture content the moist sample is weighed and then dried until the reference weight w_0 is attained. For ordinary moisture-content determination, where high precision is not required, the dry weight w_0 is obtained simply by drying the sample in an oven maintained at 103 ±2°C until constant weight is attained. This procedure reduces the sample moisture content to a low value at equilibrium with a relative vapor pressure sufficiently close to zero that the sample is assumed to have attained its dry weight w_0.

There are several objections to using the simple ovendrying method for obtaining w_0 when precise moisture contents are required. Among these are appreciable departures from zero of the relative vapor pressure in the oven, the evaporation of volatile constituents other than water from the wood during drying in the oven, and the possible effect of previous sorption history.

The relative vapor pressure $h(=p/p_0)$ in a drying oven is a function of several variables, including oven temperature, ambient room temperature and humidity, and sources of moisture within the oven. The last factor can be minimized by use of a forced-circulation oven and sufficiently long drying times so that the vapor pressure p in the oven is the same as that in the room. Under these conditions the relative vapor pressure in the oven $(p/p_0)'$ can be calculated from the equation

$$(p/p_0)' = (p/p_0)''(p_0''/p_0') \qquad (2.6)$$

where $(p/p_0)''$ and p_0'' are the ambient relative and saturated vapor pressures and p_0' is the saturated vapor pressure in the oven. For example, if the room temperature and humidity are 30°C and 60 percent, the relative vapor pressure in the oven at 103°C can be calculated from equation (2.6). Expressing p_0 in mm Hg, $(p/p_0)' = (0.60)(31.82/845.1) = 0.0226$. In the case where room temperature and humidity are 20°C and 10 percent, $(p/p_0)' = (0.10)(17.53/845.1) = 0.00207$, less than 10 percent of the value obtained in the first case. The residual moisture in the dry sample in the high temperature and humidity room situation should be in the order of 0.5 percent, whereas at the lower room or ambient conditions it should be in the order of 0.05 percent (see Figure 2.33). For example, the dry weights of thirty-eight different woods native to Venezuela were 0.42 percent lower on an average when dried in an oven in Syracuse, New York, in January compared with their ovendry weights determined in Merida, Venezuela. The difference in retained moisture at the ovendry condition is undoubtedly due to the higher ambient vapor pressure in Venezuela compared with Syracuse in the middle of winter. The conditions at the two locations approached those given in the numerical example above.

The effect of room conditions can be greatly reduced or eliminated by use of a vacuum oven. It is also possible to use lower oven temperatures in this case since the ambient relative vapor pressure $(p/p_0)''$ is essentially zero. Sometimes a strong desiccant such as phosphorus pentoxide is used instead of a vacuum because of its low vapor pressure at room temperature. However, much longer drying times are required to attain equilibrium at room temperatures compared with elevated oven temperatures.

The effect of previous sorption history on the dry weight w_0 of small wood samples vacuum-dried at room temperatures has been reported by Hergt and Christensen (1965) working with *Araucaria klinkii* in Australia. They found for example that the apparent dry weight w_0 of 40 μm thick sections was a minimum when they were dried rapidly at 27°C under vacuum directly from the water-soaked condition. However, when the same or similar samples were first conditioned to intermediate moisture contents in the hygroscopic range before being vacuum-dried, they appeared to reach equilibrium with the vacuum at somewhat higher weights than the minimum value of w_0. This indicated that a certain amount of water was retained in the samples upon vacuum drying if the samples were not completely wet just prior to rapid drying. They found that the maximum retention occurred in samples which were dried following exposure to p/p_0 from 0.4 to 0.6.

A more recent study by Christensen and Hergt (1969) showed that the percent of water retained upon vacuum drying, M_r, increased approximately linearly with the square root of time the samples were exposed to p/p_0 from 0.4 to 0.6. An equation which fits their data approximately is

$$M_r = 0.025 \sqrt{t} \qquad (2.7)$$

where t is exposure time in hours. It can be seen from equation (2.7) that exposure of thin wood samples to p/p_0 of, say, 0.5 for 100 hours results in a retained moisture content M_r after vacuum drying of 0.25 percent and after 900 hours of 0.75 percent.

Distillation Method

The evaporation of volatile components other than water during drying may cause substantial errors in the gravimetric method of measuring moisture in wood. In this case it is necessary to use a modification of the gravimetric method by heating the wood in a distillation apparatus containing a water-immiscible liquid which is a solvent for the volatile extractive compounds. The water is condensed in a reflux condenser system and separated from the solvent by

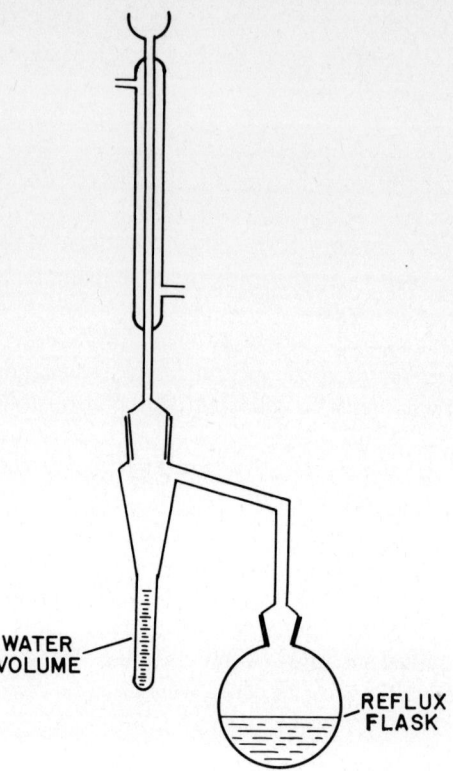

Fig. 2.6. Distillation apparatus for measuring moisture content.

means of a calibrated trap (Figure 2.6). According to Kollmann and Höckele (1962), several kinds of solvent can be used satisfactorily, including toluene, xylene, and trichloroethylene. These workers also report that the method is less exact than the Karl Fischer method discussed below, at least partially because of the inaccuracy involved in measuring the water volumetrically.

Karl Fischer Titration Method

The Karl Fischer titration method for measuring the moisture content of wood is another technique which is particularly useful for material containing volatile extractives. Kollmann and Höckele (1962) found that this technique gave the best results of several standard methods. Resch and Ecklund (1963) found that the Karl Fischer method gave lower moisture-content values than the American Society for Testing Materials (ASTM) ovendrying method for *Chamaecyparis lawsoniana* and *C. nootkatensis*, presumably because of the volatile oils in these woods.

In the Karl Fischer method the water content is measured by titration using a methanol solution of sulfur dioxide, iodine, and pyridine. At the end point of titration free iodine appears which can be detected either visually or potentiometrically, the latter method giving more precise results. Figure 2.7 shows the apparatus used by Resch and Ecklund (1963).

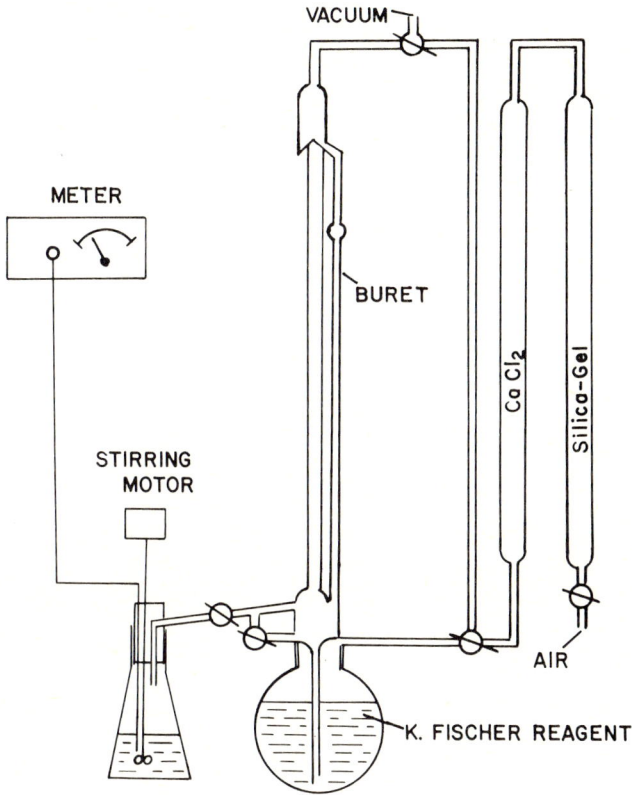

Fig. 2.7. Karl Fischer apparatus for measuring moisture content (adapted from Resch and Ecklund 1963)

Electrical Moisture Meters

Most methods for measuring wood moisture content destroy the sample to be tested and require long times to obtain meaningful measurements. However, the development of electrical methods for measuring wood moisture content has made it possible to sample the moisture content of wood nondestructively and almost instantaneously. There are two principal types of electric moisture meters, each operating according to a different principle: the resistance moisture meter, generally a direct-current (D.C.) instrument; and the dielectric or capacitance or power-loss meter, an alternating-current (A.C.) instrument. These two types of moisture meters will be discussed separately. For additional comparative information on these two types the interested reader is referred to the excellent work by James (1963).

D.C. Resistance Moisture Meter. This type of meter is essentially a D.C. megohmmeter and operates on the principle that the D.C. resistivity r of wood varies over an extremely wide range as its moisture content changes, as shown in Figure 2.8. The development of a successful resistance meter was based on the work

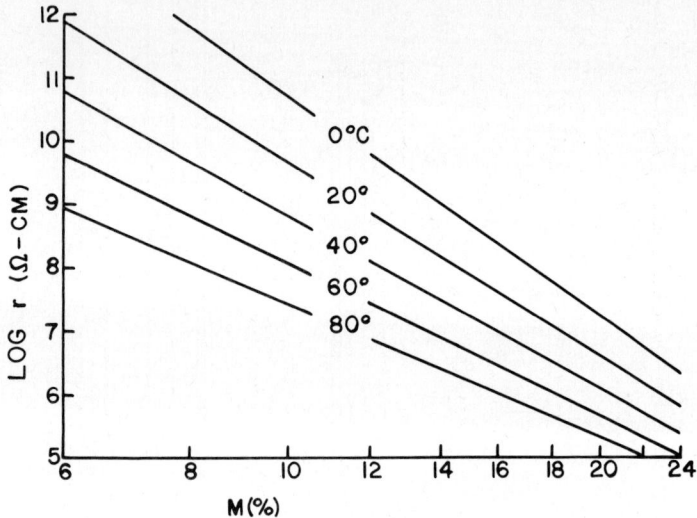

Fig. 2.8. Logarithm of the D.C. resistivity r (ohm-cm) of wood as a function of wood moisture content M (%) for several different temperatures. From *McGraw-Hill Encyclopedia of Science and Technology*, Vol. 14, figure 2. Copyright 1960 by McGraw-Hill. Used with permission of McGraw-Hill Book Company.

of Stamm (1927), who first obtained a quantitative measurement of the relationship between wood moisture content and its electrical resistivity, although Meyer and Rees (1926) also foresaw the possibility of devising an electric resistance meter for measuring wood moisture content.

Practical resistance moisture meters measure the resistance R rather than the specific resistance or resistivity r of a given wood sample. The relationship between R and r for a prismatic sample is given by

$$R = r(L/A) \qquad (2.8)$$

where L and A are the effective spacing and cross-sectional area of the electrodes. Various kinds of electrodes have been used in resistance moisture meters.

These include flat electrodes with a large surface area A and small spacing L, particularly suited for measurements on veneers, and various kinds of blade or needle electrodes which are most useful for measurements on thicker materials such as lumber. The relationship between measured resistance R and resistivity r is much more complex when needle electrode probes are used in place of the more simple geometry given by equation (2.8). For example, Figure 2.9 taken

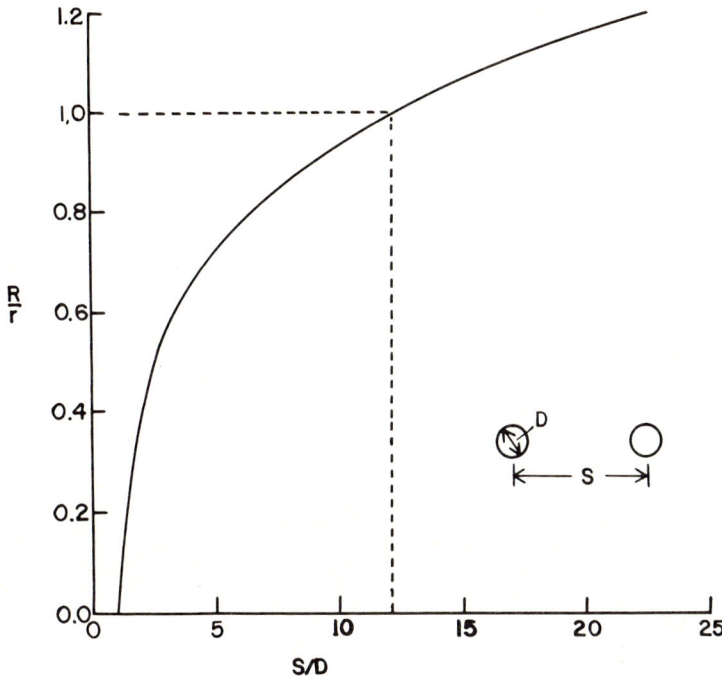

Fig. 2.9. Resistance-resistivity ratio R/r as a function of the spacing-diameter ratio S/D of the electrodes (adapted from Skaar 1964).

from Skaar (1964), shows the ratio R/r or the effective ratio L/A in equation (2.8) as a function of the ratio S/D for two cylindrical electrodes, each one centimeter in length, of diameter D in centimeters and center-to-center spacing S in centimeters. The ratio of S/D of 12.5, for example, gives an R/r ratio of 1.0. In the case of cylindrical electrodes the current flow is concentrated in the vicinity of the electrodes, and when the S/D ratio is very large the resistance measured is essentially that close to the electrodes.

It is popular practice in the United States to use two parallel pairs of needle-type electrodes in measuring the moisture content of lumber. Moisture meters are than calibrated in terms of wood moisture content using the data shown in

Table 2.1. Electrical Resistance R at Wood Moisture Content M at 80°F

M(%)	7	8	10	12	14	16	18	20	22	24
R (megohms)	22,400	4,780	630	120	33	11.2	4.6	2.14	1.10	0.60

Table 2.1. These are based on readings of resistance R against wood moisture content M for Douglas-fir at 80°F as reported in the *Wood Handbook* (1955) using two pairs of needle electrodes spaced 1.25 inches apart and driven to a depth of $\frac{5}{16}$ inch in wood of uniform moisture content. The direction of current flow is essentially parallel to the grain.

It was indicated that the resistance values in Table 2.1 are measured with the electric field parallel to the grain direction. Stamm (1960) has shown that the resistivity is two or three times as great across the grain as in the longitudinal direction, depending on the kind of wood. The resistivity is usually slightly higher tangentially than radially. These differences in resistivity with respect to structural direction in the wood are explainable on the basis of gross wood structure (Siau 1971) and partly on the basis of fibril orientation.

The resistance values given in Table 2.1 are for wood of uniform moisture content. Usually, however, the moisture content is not uniform and gradients of moisture occur within the wood. In the normal drying process, during which it is desirable to monitor the moisture content of the wood, the moisture gradient is such that the outer wood layers are drier than those in the interior. Stamm (1930) found by empirical methods that the effective resistance given by needle electrodes driven to a depth equal to one-fourth the total thickness of lumber with normal drying gradients was essentially equivalent to the resistance for lumber of uniform moisture content equivalent to the same average value. In principle it should be possible to predict the depth of penetration which will give the correct mean value for any kind of drying gradient. However this is impractical for several reasons, including the variation along the needle length of the S/D ratio and the complex three-dimensional nature of the current paths, particularly at the deepest point of penetration. Some of these factors, including the effect of the anisotropy of the electrical resistivity of wood, are discussed in Skaar (1964).

The high sensitivity of resistivity to moisture-content variations can be used to monitor the moisture distribution in lumber. For example, with the normal drying moisture gradient such as shown in Figure 2.10 the reading obtained using needle electrodes will vary markedly with depth of electrode penetration into the wood. Because of the great sensitivity to moisture content, it can be assumed to a first approximation that the indicated moisture-meter reading at any depth of penetration is equivalent to the actual moisture content at that depth. Furthermore, it is possible to detect wet pockets or regions of excessively high moisture content in otherwise satisfactorily dry wood.

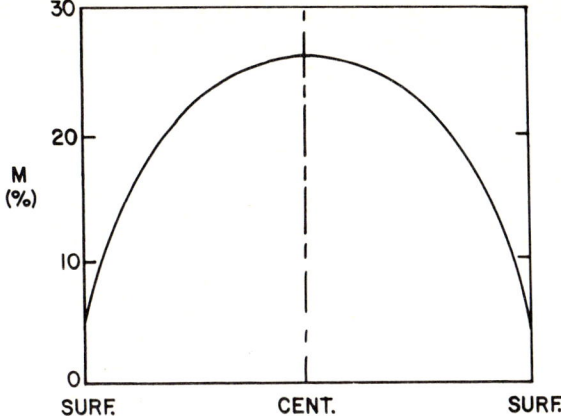

Fig. 2.10. Typical moisture distribution for wood during drying from surface to center of the wood.

When the surface layers of lumber are more moist than the interior it is necessary to use electrodes which are insulated all along their length except for the tip or point of measurement. This is because the resistance of the moist surface is so low compared with that of the interior that it acts effectively as a short-circuit for the current, and the moisture meter reads essentially the surface moisture content if the electrodes are not insulated. Some moisture meter manufacturers provide special insulated electrodes which are intended for use in measuring moisture gradients in lumber.

The reliability of resistance moisture meters is generally good in the hygroscopic moisture-content range. However, above fiber saturation their reliability is poor because of three factors. First, the variation of resistivity with moisture content is relatively small (Figure 2.11) above fiber saturation, and the sensitivity of the meters is much reduced. Second, variations in electrolyte content in the cell-cavity water between kinds of wood and even from one point in the wood to another in the same sample of wood causes large variations in resistivity. Third, time-dependent effects become more pronounced at high moisture contents. In other words, the apparent value of D.C. resistivity changes substantially with time after application of voltage (Skaar 1964).

Resistance moisture meters therefore are most useful in that part of the hygroscopic moisture range between 6 or 7 percent and 25 to 30 percent moisture content at room temperatures. When wood is drier than this the resistivity is too high to measure with ordinary commercial moisture meters, and when the wood is too wet the reliability is poor for reasons mentioned in the preceding paragraph. This is the most useful range, however, and if the meter indicates a moisture content below 6 or 7 percent the wood is too dry for most practical purposes and must be permitted to gain additional moisture. Similarly, if the

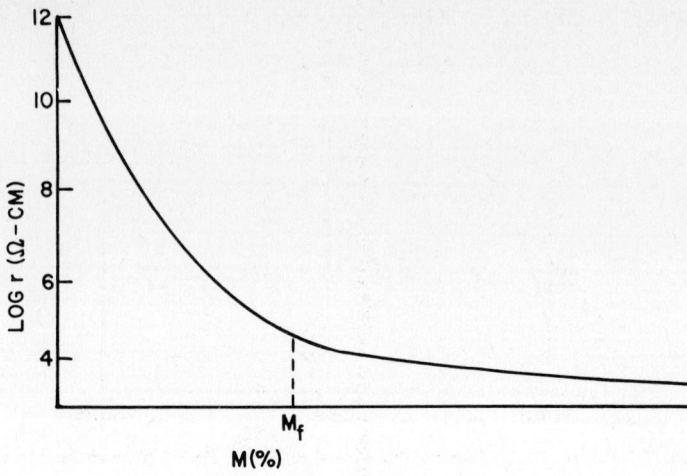

Fig. 2.11. Logarithm of D.C. resistivity r (ohm-cm) against wood moisture content M (%) above and below fiber saturation M_f.

meter indicates that it is near or above fiber saturation the wood must be further dried before being put to use.

It has already been mentioned that there are large variations in the electrical resistivities of different kinds of wood at the same moisture content above fiber saturation. There are also variations among woods with respect to their resistivities at corresponding moisture contents in the hygroscopic range. Figure 2.12, taken from data supplied by James (1963), shows this variation for a number of woods native to the United States. Because of this variation among woods it is necessary to obtain calibration data for various wood species which are then used to correct the readings obtained directly from the moisture meter. Such calibration data are normally provided by the manufacturer.

Temperature also has a large effect on the electrical resistivity of wood and therefore must be considered in interpreting moisture meter readings. Figure 2.13, taken from Lin (1965), shows that the logarithm of resistivity is a nearly linear function of the reciprocal of absolute temperature at constant wood moisture content over much of the hygroscopic moisture range. This indicates that the process of electrical conduction is an Arrhenius type of activated process. The primary conductors are believed to be charged ions, but the exact mechanism of conduction is not clear. For a discussion of the mechanism of electrical conduction the reader is referred to Lin (1965) and Venkateswaran (1971).

For practical use in making moisture-meter temperature corrections, the curves of Figure 2.14, taken from James (1963), are more useful than those of Figure 2.13. This set of correction curves is based on 70°F as the calibration temperature. For example, if the wood temperature at time of measurement is

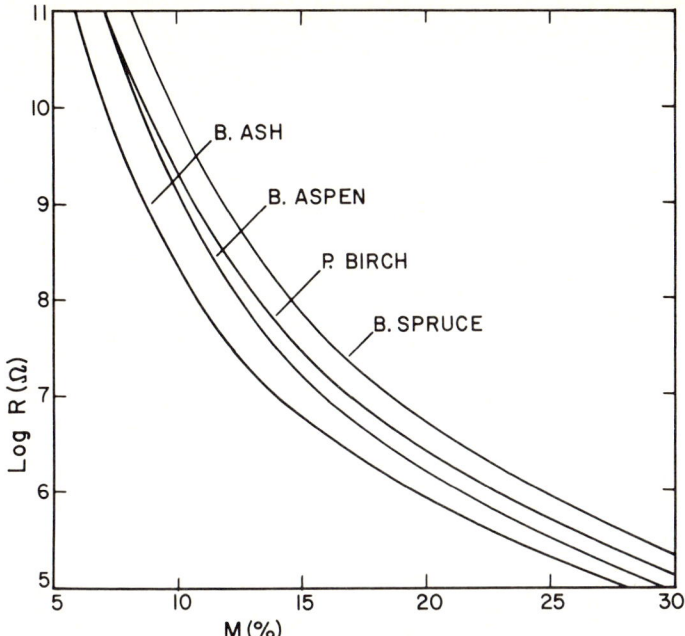

Fig. 2.12. Logarithm of D.C. resistance R (ohm) against wood moisture content M (%) for four wood species (adapted from James 1961).

120°F and the moisture meter reads 18 percent, the actual moisture content, based on the 70°F calibration temperature, is 14 percent.

The apparent decrease dM/dF in wood moisture content, as read on the resistance moisture meter per degree F increase in temperature, can be calculated at any temperature and true wood moisture content from the slope of the constant moisture-content curve in Figure 2.14. For example, at 70°F, the reference temperature, the slope of this curve increases approximately from $dM/dF = 0.040$ percent/°F at $M = 6$ percent to $dM/dF = 0.144$ percent/°F at $M = 28$ percent. In terms of dM/dC, the change in moisture-meter reading per °C (Celsius or Centigrade) are these values multiplied by 9/5, the conversion ratio between the two temperature scales. An empirical equation which gives a reasonably good approximation for the variation of dM/dF at 70°F, over the moisture range from 6 to 28 percent, based on Figure 2.14, is

$$dM/dF \approx 0.015 + 0.0047\,M. \tag{2.9}$$

A similar equation for dM/dC at the same temperature (21°C) is

$$dM/dC \approx 0.027 + 0.0085\,M. \tag{2.10}$$

Values of dM/dC at 21°C are given by Skaar (1964) for the heartwood and

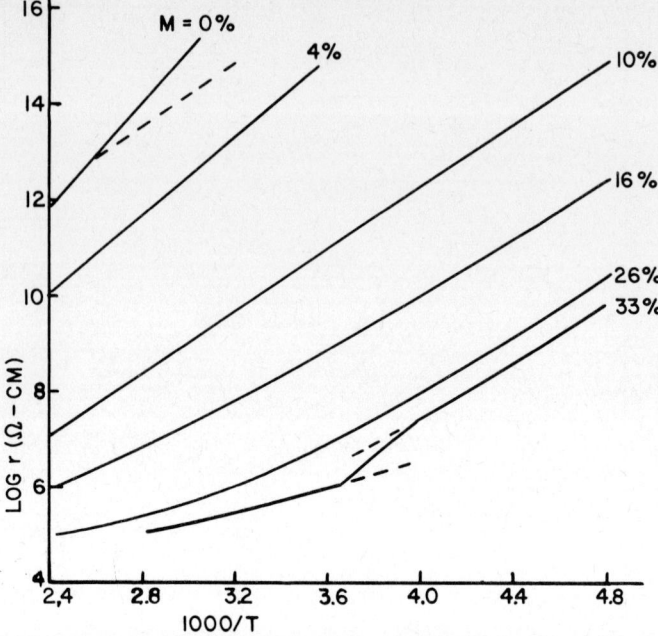

Fig. 2.13. Logarithm of D.C. resistivity r (ohm-cm) against the reciprocal of absolute temperature T (°K) at several moisture contents M (%) for yellow poplar (adapted from Lin 1965).

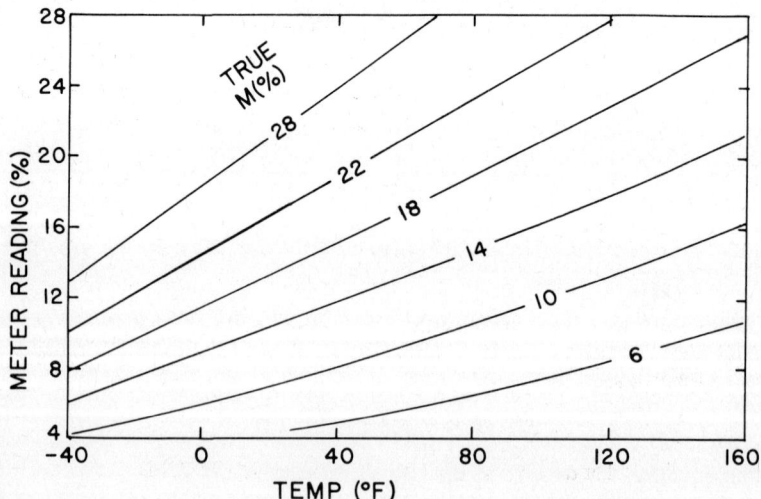

Fig. 2.14. D.C. resistance moisture-meter temperature calibration curves for different true wood moisture contents (adapted from James 1963).

Table 2.2. Values of dM/dC (%/°C) at 21°C for Heartwood H and Sapwood S
of 5 U.S. Woods as Reported by Skaar (1964)
Compared with Those Calculated by Use of Equation (2.10)

| M | Doug. Fir | | Yell. Poplar | | Bl. Cherry | | Yell. Birch | | Pond. Pine | | Mean of | Equation (2.10) |
(%)	H	S	H	S	H	S	H	S	H	S	All Species	
12	.112	.136	.133	.133	.113	.134	.139	.144	.128	.134	.131	.129
18	.129	.157	.158	.155	.140	.187	.168	.158	.146	.153	.155	.180
24	.122	.163	.167	.162	.144	.190	.180	.171	.147	.155	.160	.231

sapwood of five North American woods at moisture contents of 12, 18, and 24 percent. These are listed in Table 2.2 together with the values calculated by equation (2.10) at these same moisture contents. It is clear from the table that there are substantial differences among species with respect to the magnitude of the temperature effect dM/dC and also in some cases between sapwood and heartwood of the same species. Furthermore, the values of dM/dC obtained from the curves of James (1963) increase more rapidly with moisture content than do the values given for the individual woods. The practical conclusion that may be drawn from the table is that individual species calibrations are required not only for the effect of moisture content on moisture-meter readings but probably also for the temperature corrections. Furthermore, it is not really known how large the variations are within species, whether in the same or different trees, compared with those between species. For precise use of resistance moisture meters it may therefore be necessary to have calibrations for the material which is currently being studied. However, for practical field work the general information currently available is probably adequate.

A.C. Dielectric Moisture Meters. Moisture meters of the dielectric type are operated with alternating currents of sinusoidal waveform and usually at radio frequencies. In recent years the frequency range in some meters has been extended into the microwave region. Dielectric moisture meters generally operate on the principle that the dielectric constant* ϵ and the loss tangent, tan δ, of wood increase markedly with wood moisture content at a given frequency. Some meters, generally known as capacitance meters, measure only the variation in dielectric constant ϵ of wood with moisture content. The more common type, known as power-loss meters, measure the combined effects of variation in both dielectric constant and loss tangent with moisture content. In order to understand dielectric moisture meters it is necessary to review briefly some of the principles involved in dielectric measurements and the principal factors in wood which affect its dielectric properties.

---

*In more modern terminology the dielectric constant is considered to consist of two parts, an "in-phase" component ϵ' and an "out-of-phase" component ϵ''. The definition we use here is essentially that of the component ϵ'. The loss tangent, tan $\delta = \epsilon''/\epsilon'$.

The dielectric constant of a material is a measure of the electric polarizability of the constituents per unit volume of the material. That is, when a D.C. voltage is applied to a material such as wood, the positively charged components in the cell wall are attracted toward the negative electrode and the negatively charged particles toward the positive electrode. In other words, the charges tend to become polarized in an electric field. If an A.C. voltage is applied the charges also tend to become polarized in the same way at any instant. In this case, however, the direction of polarization undergoes a complete change during each cycle. The total amount of electrical polarization that takes place during each cycle is a measure of the dielectric constant of the material. In general the amount of polarization, and therefore the dielectric constant, is dependent on the period or the frequency of alternation of the electric field.

The loss tangent, on the other hand, is a measure of the fraction of energy lost during each cycle due to friction. The loss tangent is also dependent on the frequency of the electric field.

A.C. measurements of dielectric constant and loss tangent have been made over a wide frequency range (Figure 2.15). These measurements made by Trapp and Pungs (1956) show that ϵ, for wood at constant moisture content and temperature, decreases as the measuring frequency increases. It is also clear that the dielectric constant increases with increasing wood moisture content at any given frequency. The loss tangent behaves in a more complex manner, showing peaks

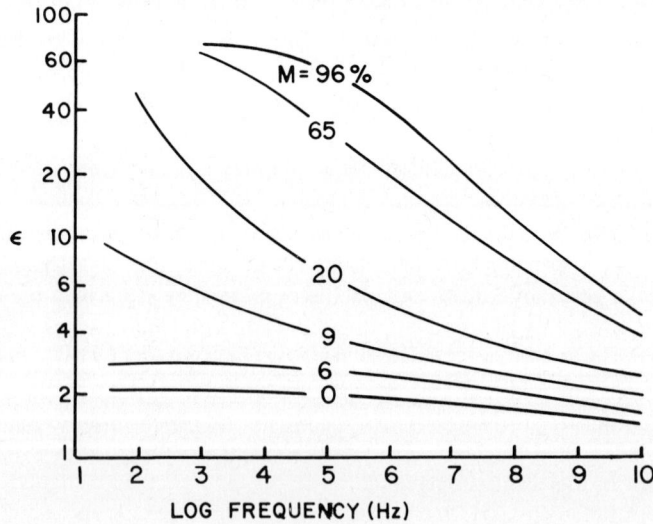

Fig. 2.15. Dielectric constant ϵ as a function of the logarithm of frequency (Hz) at 20°C for European spruce at several different moisture contents (adapted from Trapp and Pungs 1956).

at certain frequencies, corresponding to those at which the dielectric constant is changing most rapidly with frequency.

The frequencies at which the loss-tangent peaks occur are characteristic of the wood at a given moisture content and temperature. The time period associated with any given frequency is the reciprocal of the frequency. The period of the peak frequency is proportional to a certain time constant T which may be defined loosely as the time required for a particular kind of polarization to take place after a voltage has been applied to the wood.

There are two characteristic peak frequencies corresponding to two electrical time constants observable in wood. The peak which occurs at the lower frequencies, usually at audio or power frequencies, is believed to result from interfacial polarization caused by the accumulation of charged ions at interfaces within the cell wall of wood. The exact nature of these interfaces is not known; they may be interfaces between cells or between the middle lamella and the secondary wall or between crystalline and amorphous regions. Figure 2.16, taken

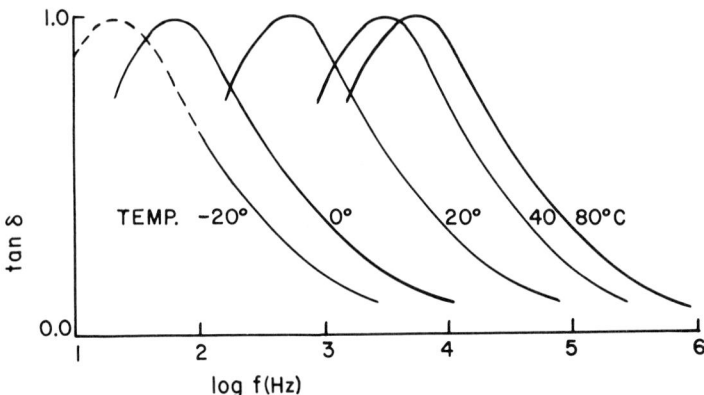

Fig. 2.16. Loss tangent (tan δ) against the logarithm of frequency (*Hz*) for wood at 15 percent moisture content at several temperatures (adapted from Tsutsumi and Watanabe 1965).

from Tsutsumi and Watanabe (1965), shows such peaks for Japanese Sakura wood at 15 percent moisture content.

The second type of polarization, found in the radio-frequency range or higher, is dipole polarization. It is caused by the rotation of permanent dipoles such as hydroxyl groups or hygroscopic water in the cell wall. Dipoles rotate in the alternating electric field and contribute to the dielectric constant at frequencies sufficiently low to follow the changes in the A.C. voltage. The time constant for dipole rotation is in the order of microseconds (10^{-6}) to picoseconds (10^{-12}), depending on the moisture content and temperature of the wood. Figure 2.17

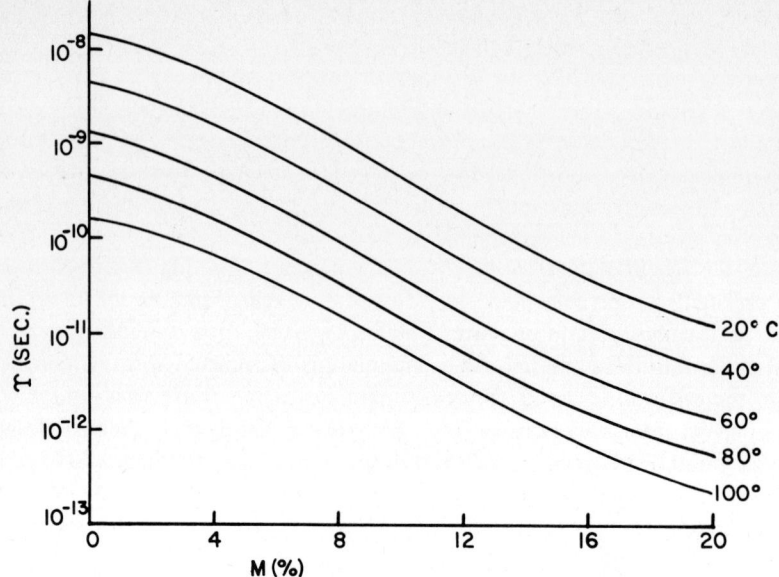

Fig. 2.17. Dipole time constant τ (sec.) as a function of wood moisture content M (%) for different temperatures °C.

shows the variation in the dipole time constant for wood with moisture content and temperature.

The dielectric properties of most interest in dielectric moisture meters are those in the radio-frequency and microwave region since these meters usually operate in this range. Because of the complex interaction of ϵ and tan δ with frequency it is best in a practical meter to keep the frequency constant or nearly so. With this is mind it is possible to show how wood factors such as moisture content and density affect these properties.

Most dielectric moisture meters operate at radio frequencies where dipole polarization is most important. At a given radio frequency, say one in the 1 to 10 Megahertz (10^6 Hertz) range, the dielectric constant is markedly affected by both wood moisture content and density. The dielectric constant of the dry cell wall at these frequencies is in the order of 4 (Skaar 1948) while that of liquid water is about 80. Therefore, as the wood moisture content increases the wood dielectric constant increases. This increase is nonlinear in the hygroscopic range but apparently linear above fiber saturation (Figure 2.18). The reason for the nonlinearity in the hygroscopic range is believed to be associated primarily with the restricted freedom of rotation of the sorbed water at these moisture contents. As the moisture content increases, the added water molecules have greater free-dom of rotation and therefore make a greater contribution per molecule to the dielectric constant than those added at lower moisture contents. Above fiber

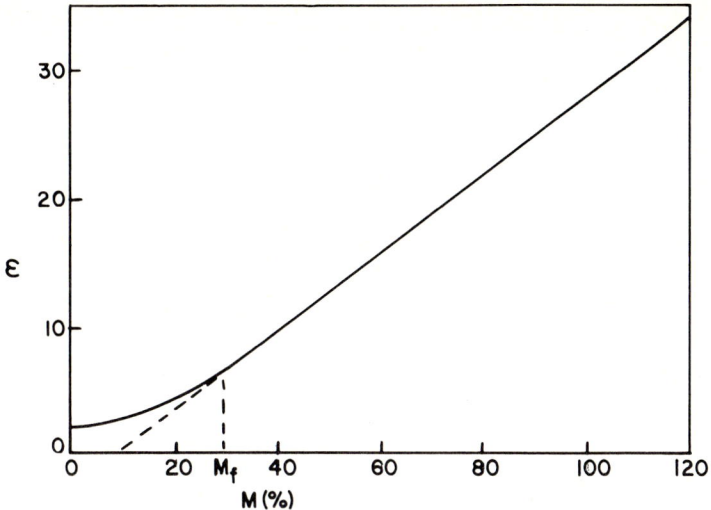

Fig. 2.18. Dielectric constant ϵ against wood moisture content M (%) for buckeye at a frequency of 2 MHz (adapted from Skaar 1948).

saturation the added water has essentially the same effective dielectric constant as liquid water and the curve tends to become linear, as indicated in Figure 2.18.

Since wood consists of both cell-wall material and void space (the dielectric constant of which is essentially unity), it is anticipated that the dielectric constant of wood should increase with density at a given moisture content. The experimental data shown in Figure 2.19 indicate that the increase is somewhat nonlinear with increasing density for the perpendicular-to-grain dielectric constant. Increasing temperature has a similar effect on the dielectric constant as an increase in moisture content at a given frequency, as Figure 2.20, taken from Nanassy (1964), illustrates.

The loss tangent varies in a complex manner with moisture content, depending on the frequency of measurement. Generally it increases with moisture content, but apparent anomalies occur as shown in Figure 2.21, taken from Hearmon and Burcham (1954).

The capacitance moisture meter measures the electrical capacitance C between two electrodes in which wood is the dielectric material. The capacitance C is proportional to the dielectric constant ϵ, or

$$C = k\epsilon \qquad (2.11)$$

where k is the proportionality coefficient, determined by the nature of the electrodes and the kind of electrical coupling to the wood. The coupling coefficient may be constant for a given electrode configuration, in which case the ratio C/ϵ is constant, or it may be a function of ϵ. For example, the latter would be

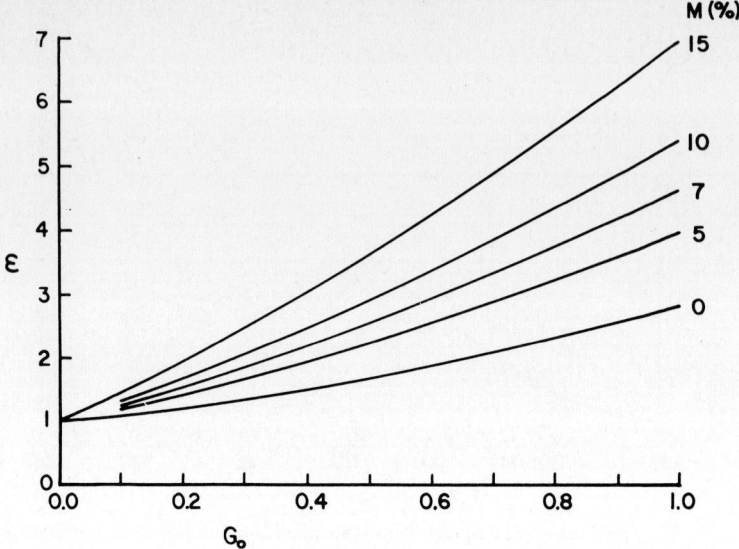

Fig. 2.19. Dielectric constant ϵ against dry wood density for several different wood moisture contents M (%) (adapted from Uyemura 1960).

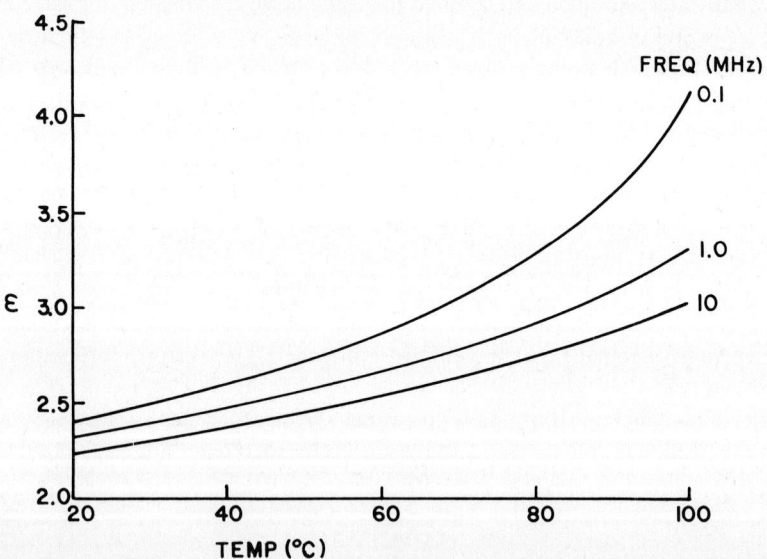

Fig. 2.20. Dielectric constant ϵ against temperature (°C) for dry wood at three different frequencies (*MHz*) (adapted from Nanassy 1964). Reproduced by permission of the National Research Council of Canada from the *Canadian Journal of Physics*, 42 (1964), pp. 1270–8.

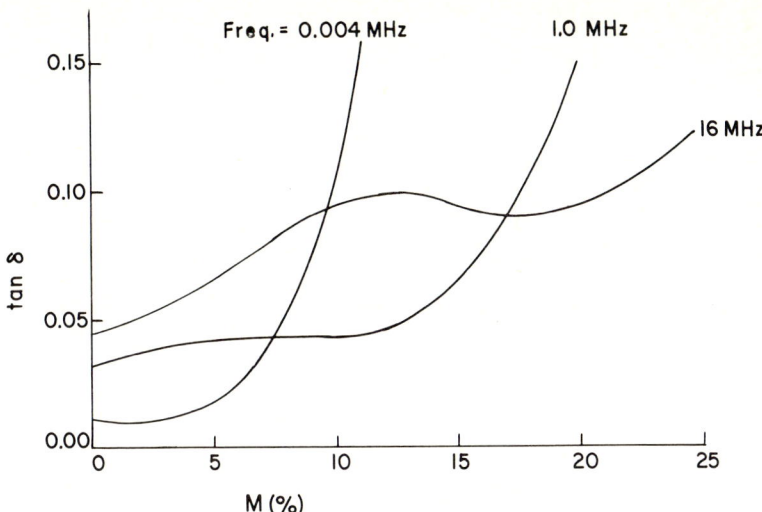

Fig. 2.21. Loss tangent (tan δ) against wood moisture content M (%) for three different frequencies (*MHz*) (adapted from Hearmon and Burcham 1954).

true if the coupling of the electrodes to the wood is through an air gap. The electrode configuration is an important design consideration and must be a compromise between practical use factors and ideal considerations.

It is clear from equation (2.11) and Figure 2.19 that the measured capacitance is dependent not only on the wood's moisture content but also on its density. Therefore it is necessary to have a calibration curve for a given species based on the mean density of the wood. Furthermore, the appreciable density variations which occur within a given species also will cause similar variations in the measured moisture content. In principle, the capacitance meter can be used at wood moisture contents over the entire range from dry to complete saturation. However, because of the excessively high loss factor at high moisture contents it is difficult to make meaningful measurements in this range unless special provisions are made for air coupling between the electrodes and the wood.

The power-loss meter, the most common type of dielectric moisture meter, measures the electrical power absorbed by a given wood sample when electrodes of a standard configuration and electrode coupling are used. The power loss P in the sample is proportional to the dielectric constant ϵ and also to the loss tangent, tan δ, or

$$P = (K\epsilon)(\tan \delta) \qquad (2.12)$$

where K is the coefficient of proportionality, determined by factors similar to those which determine k in equation (2.11)

It is clear from the equation that the value of P is dependent on ϵ as well as tan δ and that similar electrode factors are involved as with the capacitance meters. It is therefore necessary to calibrate the power-loss meter for different kinds of wood, particularly with respect to density.

It is anticipated that the A.C. dielectric meters would be sensitive to temperature since ϵ and tan δ both change with temperature. James (1968) has made temperature-correction data for typical radio-frequency moisture meters given in terms of scale readings. Figure 2.22 shows scale reading curves for one of

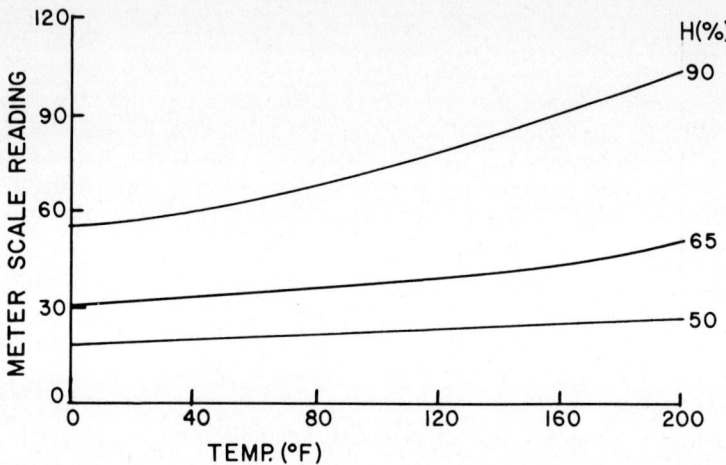

Fig. 2.22. Temperature correction curves for a radio-frequency moisture meter for wood at three different equilibrium relative humidities H (adapted from James 1968).

these meters as functions of temperature for wood conditioned to equilibrium with three different relative humidities H. It is clear from the correction curves for the meter, which is believed to be of the capacitance type, that the indicated moisture-meter reading is too high at higher temperatures and too low at lower temperatures, compared with the 80°F standard temperature. This indicates that increasing temperature has the same effect on capacitance of dielectric constant as increasing moisture content. In this respect the correction curves for the power-loss meter should be more complex, however, because of the interaction of loss tangent and dielectric constant.

The grain orientation also affects capacitance or power-loss moisture meters because both ϵ and tan δ are higher along the grain than in the transverse direction (Skaar 1948).

Microwave moisture meters are a recent development of the power-loss radio-frequency moisture meters. They differ from the ordinary radio-frequency wood moisture meter in two respects.

First, the frequencies are much higher, since microwave frequencies are generally considered to range from frequencies of one Gigahertz (10^9 Hertz) upward to 100 Gigahertz. The term "microwave" is used because the wavelength in the Gigahertz (GHz) region is small compared with those of ordinary radio frequencies, ranging from 30 cm at 1 GHz to 0.3 cm at 100 GHz. Because of the short wavelengths involved, special techniques are used for generating these high frequencies and also for transmitting and coupling them to the wood to be measured. Discussion of these techniques is outside the scope of this book, and reference is made to James and Hamill (1965) for more information on the techniques of measurement at microwave frequencies.

A second distinction between microwave moisture meters and classical radio-frequency power-loss meters is that the microwave meters may operate at one of the characteristic frequencies at which maximum power absorption by the water molecule occurs. At this frequency, 22.235 GHz (1.35 cm wavelength), according to Walker (1964), a resonance occurs in the free-water molecule because of rotation of the molecule. This characteristic absorption frequency seems to apply also for at least part of the water sorbed by hygroscopic materials such as paper and wood. For example, Walker (1964) indicates an attenuation or absorption peak at this frequency for paper containing 12 percent water. There is indication, as is true for radio frequencies, that the power loss or attenuation increases nonlinearly with moisture content in the hygroscopic range for paper, according to Busker (1968), but linearly above this range (Figure 2.23).

It is not known by the author if commercial microwave moisture meters operate at the characteristic molecular rotational resonance frequency of 22.235 GHz, although this seems probable. It is also possible to use other frequencies

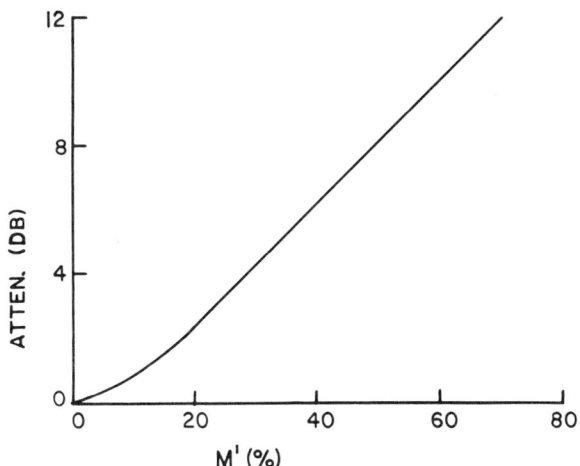

Fig. 2.23. Power attenuation in decibels (*DB*) against moisture content *M'* (%, wet basis) of paper (adapted from Busker 1958).

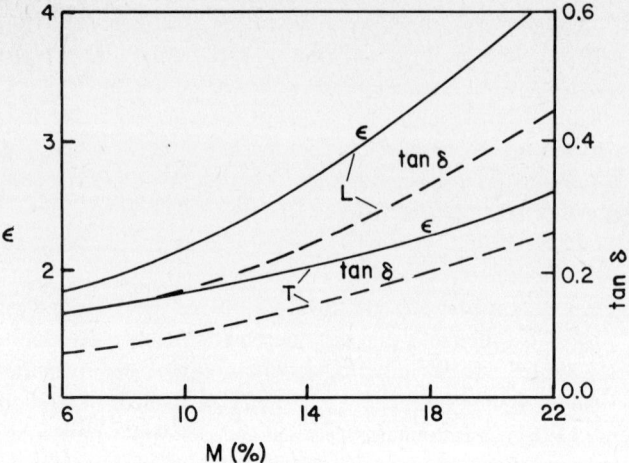

Fig. 2.24. Dielectric constant ϵ and loss tangent, tan δ, of Douglas-fir against wood mois-ture content M (%) for both longitudinal (L) and transverse (T) field orientation at 8.53 GHz (adapted from James and Hamill 1965).

since the dielectric constant and loss tangent generally increase with moisture content at other microwave frequencies. For example, Figure 2.24 adapted from James and Hamill (1965), shows the increase in both the dielectric constant and loss tangent of Douglas-fir, with moisture content in the hygroscopic range for longitudinal and tangential orientations of the electric field, at 8.53 GHz. It is clear that the shapes of the curves are generally similar to those in the Megahertz range, although the dielectric constant is considerably lower, as anticipated. It should also be noted that both dielectric constant and loss tangent are higher for the longitudinal or parallel-to-grain orientation, as is true at the lower radio frequencies. The tangential values for this wood were somewhat higher than the radial values.

Lowery and Kotok (1968) have evaluated the performance of a commercial microwave wood moisture meter on five softwood species from the Rocky Mountain region of the United States over the moisture range from the green condition to ovendry. They concluded that the moisture content could be measured with a relatively high degree of accuracy. However, they caution that the readings are sensitive to wood density, as would be expected since the dielectric absorption should be proportional to the quantity of water per unit volume of the wood, as is the case with the standard radio-frequency meters. They also indicate that wood temperature is a factor which should be considered, in agreement with the findings of Busker (1968).

Salamon (1971) has recently reviewed the characteristics of three of the common types of electric moisture meters used in British Columbia, including a D.C. resistance meter and radio-frequency meters of both the capacitance and power-loss types. He concluded that the D.C. resistance meter and the radio-frequency meters gave similar accuracy in measuring the moisture content of kiln-dried and surfaced samples of white spruce and lodge-pole pine of $1\frac{5}{8}$-inch thickness, compared with ovendry moisture determinations. He indicated that the tolerance levels of precision were ±1.5 percent for lodge-pole pine and ±3.0 percent for white spruce.

Miscellaneous Methods

One of the applications of nuclear energy to industry has been in the development of the "beta-gauge" for measuring the density and/or moisture content of paper and wood. Beta particles or rays are fast electrons emitted from the nuclei of certain isotopes during radio-active decay. When they pass through matter, such as wood, they are absorbed approximately as an exponential function of the mass of material through which they pass. Therefore, if a beta source is placed against one surface of a wood sample with a detector of beta radiation on the opposite surface, the rate or number of beta particles per unit of time which reaches the detector is essentially proportional to the wood density for a given thickness.

Noack and Kleuters (1960) used this principle to measure the moisture content of wood veneers during the drying process and concluded that it might be a practical method for monitoring the moisture content of veneers or wood chips during the manufacturing or drying processes. One basic limitation in the use of beta particles for this purpose is their high absorptivity in solid materials such as wood, thus limiting the upper range of thicknesses to that of veneers. It is of interest to note that this technique has been used to monitor density variations through the annual rings of wood as a tool for growth-quality studies (Phillips *et al.* 1962).

Another application of nuclear radiation techniques to moisture-content measurement involves the use of two different kinds of radiation simultaneously. A "fast" neutron generator is the source of high-energy neutrons which are directed into the material to be measured (Figure 2.25). Some of these fast neu-

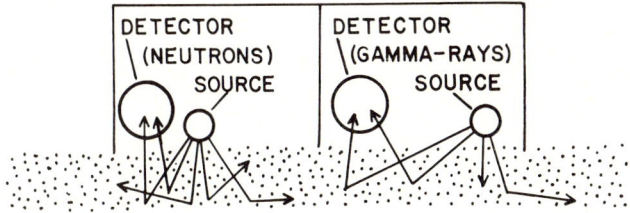

Fig. 2.25. Schematic diagram of a nuclear gage for measuring moisture content of bulk materials such as wood chips (adapted by courtesy of Nuclear-Chicago Corp.)

trons lose much of their energy in the material and become slow or "moderated" neutrons as they are reflected back to a slow-neutron detector. Hydrogen is a particularly effective neutron moderator, and since each molecule of water contains two atoms of hydrogen, the amount of moisture in the material has a large effect on the number of moderated neutrons received by the detector. Dry wood also contains hydrogen, of course, and it is therefore necessary to calibrate an instrument for this background.

It is also necessary to determine the density of the material in order to measure its moisture content on a weight basis since the neutron radiation system measures the amount of water per unit volume of material. This is accomplished by means of a gamma radiation and detection system. A beam of gamma rays is directed into the material; the intensity of radiation reflected or transmitted through the material to the gamma detector is inversely proportional to the density of the material (Figure 2.25). The combined output data obtained from the two radiation systems can be combined to give the moisture content on a weight basis. This system is readily adaptable to measuring granular materials such as wood chips in a continuous process such as particleboard manufacture.

Another indirect method of estimating wood moisture content is to measure its equilibrium relative vapor pressure h. If the sorption isotherm is known this can be related to the wood moisture content.

It is generally more important to know the equilibrium relative vapor pressure of wood during use than the moisture content itself because the variations in atmospheric vapor pressure p compared with the vapor pressure of the wood determine whether the wood is gaining or losing moisture and therefore whether it is swelling or shrinking at any given moment.

One method for measuring the internal relative vapor pressure in a wood sample is to drill a hole in the sample and immediately insert a humidity-sensing element into the hole, then seal the hole to reduce moisture loss. Usually the humidity-sensing element, called a wood hygrometer in this case, consists of a hygro-expansion element such as a hair element instrument or similar device (Kollmann and Côté 1968). It is also possible to use a miniature electric hygrometer of the lithium chloride type which has the advantage of rapid response time. Duff (1966) has used electric hygrometers made from small wood probes (Figure 2.26) which have been used to indicate vapor pressures in small holes in the wood, as has Jimènez (1967). When using such electrical devices sufficient time must be permitted for the probes to reach temperature equilibrium with their surroundings. This may be important because of the heat generated when the probe takes up moisture and vice versa.

The equilibrium relative vapor pressure of the surface of a wood sample can be estimated by placing the sample into a sealed container and measuring the equilibrium relative vapor pressure in the container. Any of the hygrometer types mentioned above could be used to measure the equilibrium conditions. The con-

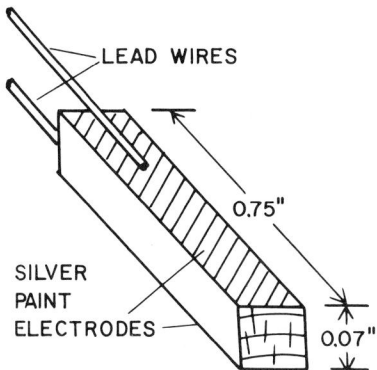

Fig. 2.26. Wood moisture probe (adapted from Duff 1966).

tainer's volume should be small compared with that of the wood, and the system should be at temperature equilibrium.

Lowery (1971) has measured the vapor pressure and temperature changes within Engelmann spruce samples during drying in a heated oven. He used a mercury manometer connected to a 2 cc air chamber in the wood and a thermocouple to make his measurements. This technique is not useful at low temperatures because the water-vapor pressure of the wood is only a small fraction of the total air pressure in this case.

Water Content of Green Wood

Water is a fundamental constituent of the living tree, as it is of all living organisms. The xylem or wood of the tree, in addition to giving mechanical support and providing food-storage capacity, also provides the passageways by which water is carried from the roots to the foliage. The active passageways for water transport are restricted to the sapwood, generally to the most recently formed sapwood.

The driving mechanism by which water is transported from the roots to the leaves is believed to be predominantly by the diffusion pressure deficit caused by transpiration from the leaves. According to this mechanism the reduced vapor pressure of the water in the leaves is caused by evaporation into the drier atmosphere. Associated with this reduced vapor pressure is a liquid tension in the water in the cells from which evaporation is taking place. This tension pulls on the liquid water in the water columns, which are more or less continuous to the tree roots, and draws water upward in the sapwood. Thus it replaces the water which has transpired as well as that used up in photosynthesis.

When a tree is cut, the process of water transport is interrupted. The moisture in the tree at the time it is felled is called the green moisture content. Studies of the moisture content of green wood show that considerable variation exists among kinds of trees, between heartwood and sapwood in the same tree, and even between logs cut from different heights of the tree. There are also seasonal variations and variations among trees of the same species.

Tables 2.3 and 2.4, taken from Peck (1953), show the average green moisture content of sapwood and heartwood of a number of United States woods, based on their ovendry weights. It can be seen from Tables 2.3 and 2.4 that there is considerable variation among trees with respect to their green moisture contents, which range from 30 percent for Douglas-fir heartwood to 249 percent for western red cedar sapwood. Furthermore, sapwood generally contains much more moisture (mean 148.9 percent) than heartwood (55.4 percent) in the case of softwoods, while there appears to be no consistent difference (mean of

Table 2.3. Average Green Moisture Contents of Softwoods Grown in USA (adapted from Peck 1953)

Species	Moisture Content, %	
	Heartwood	Sapwood
Baldcypress	121	171
Cedar, Alaska	32	166
Cedar, incense	40	213
Cedar, Port-Orford	50	98
Cedar, western red	58	249
Douglas-fir, coast type	37	115
Douglas-fir, intermediate type	34	154
Douglas-fir, Rocky Mountain type	30	112
Fir, grand	91	136
Fir, noble	34	115
Fir, Pacific silver	55	164
Fir, white	98	160
Hemlock, eastern	97	119
Hemlock, western	85	170
Larch	54	119
Pine, lodgepole	41	120
Pine, ponderosa	40	148
Pine, red	32	134
Pine, loblolly	33	110
Pine, longleaf	31	106
Pine, shortleaf	32	122
Pine, sugar	98	219
Pine, western white	62	148
Redwood, old-growth	86	210
Spruce, eastern	34	128
Spruce, Engelmann	51	173
Spruce, Sitka	41	142
Mean for 27 Softwoods	55.4	148.9

Table 2.4. Average Green Moisture Contents of Hardwoods Grown
in USA (adapted from Peck 1953)

Species	Moisture Content, %	
	Heartwood	Sapwood
Apple	81	74
Ash, white	46	44
Aspen (quaking and bigtooth)	95	113
Basswood	81	133
Beech	55	72
Birch, paper	89	72
Birch, sweet	75	70
Birch, yellow	74	72
Cottonwood, black	162	146
Elm, American	95	92
Elm, cedar	66	61
Elm, rock	44	57
Hackberry	61	65
Hickory, bitternut	80	54
Hickory, mockernut	70	52
Hickory, pignut	71	49
Hickory, red	69	52
Hickory, sand	68	50
Hickory, water	97	62
Magnolia	80	104
Maple, silver (soft)	58	97
Maple, sugar (hard)	65	72
Oak, California black	76	75
Oak, northern red	80	69
Oak, southern red	83	75
Oak, southern swamp	79	66
Oak, white	64	78
Sweetgum	79	137
Sycamore, American	114	130
Tupelo, black	87	115
Tupelo, swamp	101	108
Tupelo, water	150	116
Walnut, black	90	73
Yellow poplar	83	106
Mean for 34 Hardwoods	81.4	82.7

81.4 percent for sapwood and 82.7 percent for heartwood) in the case of hardwoods.

Some variation in green moisture content occurs with the season of the year. In some cases it is appreciable but in others not. For example, according to Clark and Gibbs (1957), the mean moisture content of yellow birch trees grown in New Brunswick, Canada, varied from a peak of 90 percent in late April to 50 percent in September. On the other hand, sweetgum growing in Louisiana showed mean moisture contents of 105, 121, 92, and 98 percent in samples taken in April, July, October, and January, according to Henderson and Choong (1968). In temperate climates green moisture content generally declines during the summer and reaches a minimum in autumn. Such variation is not pronounced in most species, however, according to the curves shown in Figure 2.27, based on data supplied by Peck (1953).

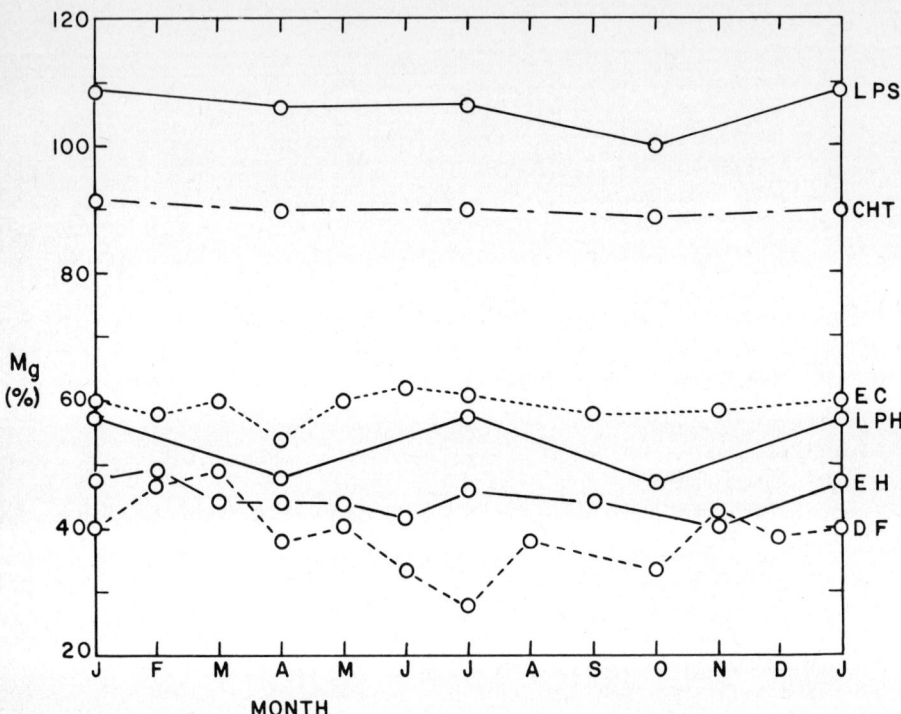

Fig. 2.27. Moisture content M_g (%) variation in green wood throughout the year from January of one year to the next, from data of Peck (1953). The woods shown are loblolly pine sapwood (LPS) and heartwood (LPH), chestnut (CHT), the means of five European conifers (EC) and of twenty-four European hardwoods (EH), and Douglas-fir (DF).

Chalk and Bigg (1956) have shown in their study of the comparative moisture distribution in trees of Sitka spruce and Douglas-fir grown in England that it is more instructive to express moisture content in terms of a fraction of the saturated capacity of the wood rather than in terms of dry weight. Their method eliminates the differences in density of the wood from different parts of the tree and is particularly useful in studies of tree moisture-content variation from the viewpoint of the physiology of moisture movement in living trees. From the viewpoint of utilization of wood, however, the moisture content based on dry weight of the wood is more useful and is therefore used in this book.

Equilibrium Moisture Content of Wood

It is clear from Tables 2.3 and 2.4 that wood in the living tree always has a moisture content of 30 percent or higher. When green wood is exposed to atmospheric conditions, however, it loses moisture to the atmosphere until it comes to a low enough moisture content that it is at equilibrium with atmospheric moisture. This moisture content, called the equilibrium moisture content (EMC), varies with the relative humidity of the atmosphere surrounding it. It also varies somewhat among different wood species, between heartwood and sapwood of the same species, and with the extractive content of the wood. The EMC is affected by temperature, mechanical stress, and by the previous exposure history of the wood. The detailed effects of some of these factors will be discussed next.

Effect of Relative Humidity and History

The single most important factor which influences the EMC of wood is the percent relative humidity H or relative vapor pressure h of the surrounding atmosphere. This relationship is generally sigmoid as is shown in Figure 2.28, taken from Spalt (1958).

It should be noted that there are essentially three curves in Figure 2.28. The

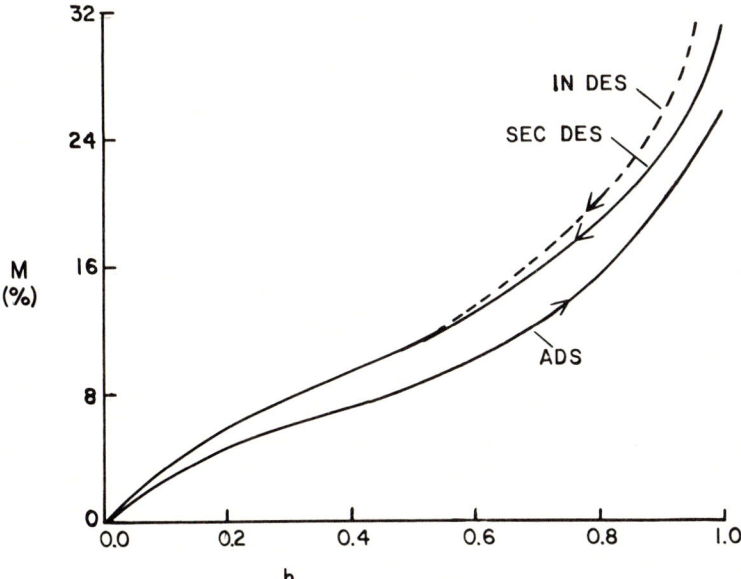

Fig. 2.28. Initial desorption (IN DES), adsorption (ADS) and secondary desorption (SEC DES) isotherms for Douglas-fir (adapted from Spalt 1958).

highest curve, the initial desorption curve obtained from green wood which has not previously been dried, gives the highest EMC at a given humidity. It was obtained by exposing the green wood to successively lower humidities from the green condition to the vacuum-dry condition, permitting sufficient time at each point to reach essential weight equilibrium within the sensitivity of the weighing balance system.

The lowest curve is the adsorption curve obtained when the sample has been vacuum dried and then allowed to regain moisture at the same humidities to which it was exposed during initial desorption, except that now the successive sequence of humidities is increasing. The intermediate or second desorption curve is obtained upon subsequent desorption after the sample has been allowed to come to equilibrium with saturated water-vapor atmosphere. It corresponds with the initial desorption curve for humidities of approximately 50 percent or lower. The last two curves are essentially reproducible if the vapor-pressure cycle is repeated, according to Spalt (1958).

The hysteresis coefficient A/D may be defined as the ratio of the EMC for adsorption to that at desorption for any given relative humidity. When the complete adsorption-desorption cycle is used, it ranges about 0.8 to 0.9, depending upon the wood and on the temperature. Figure 2.29 shows two curves for the

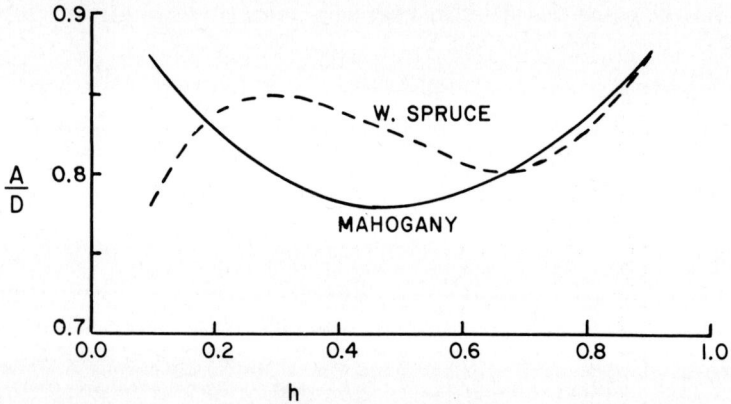

Fig. 2.29. Hysteresis ratio A/D against relative vapor pressure h for mahogany (Wangaard et al. 1967) and for white spruce (Spalt 1957).

hysteresis ratio A/D as a function of p/p_0 obtained from different sources. The mean value varies with type of wood. For example, the mean values of A/D for twelve woods, including five native to the United States and seven from foreign tropical countries, ranged from 0.80 to 0.88 with a mean of 0.84, according to Higgins (1957). His measurements were made in air over saturated salt solutions at equilibrium with p/p_0 of 0.21, 0.43, 0.81, and 0.93 at 90°F.

Spalt (1958) also found considerable variation in the mean ratio of A/D for eight North American softwoods and eight hardwoods (one from North America and the others of tropical foreign origin) at $90°F$ using a vacuum sorption apparatus over the entire hygroscopic range. In this case Spalt calculated the A/D ratio in terms of the ratio of the area under the total adsorption isotherm to that under the total desorption isotherm from $0 < p/p_0 < 1.0$. The softwood values ranged from 0.790 to 0.849 with a mean of 0.828 and the hardwood values from 0.785 to 0.844 with a mean of 0.812. Statistically the two groups were not significantly different from each other in this respect.

An unpublished study was made by this author (1967) of the EMC at $40°C$ and 76 percent relative humidity of thirty-eight different tropical hardwoods from the Guayana region of Venezuela. Two end-matched samples were obtained (0.8 cm longitudinal and 5×5 cm transverse dimensions) from green wood of each species. One sample was dried in a $103°C$ oven overnight, and then both samples were exposed to the controlled conditions in a forced-air circulation cabinet at $40° \pm 1°C$ and 76 ± 2 percent. Samples were weighed daily until equilibrium was essentially attained in all the samples after three weeks. Fifteen of the samples appeared to be sapwood and twenty-four were heartwood. Table 2.5 gives a summary of the data. The mean A/D ratio for the fif-

Table 2.5. Summary of EMC Data for 39 Different Tropical Hardwoods from the Guayana Region of Venezuela, Conditioned to $40° \pm 1°C$ and $76 \pm 2\%$ Humidity

Source	No. Species	Ads EMC (%)	Des. EMC (%)	A/D Ratio
Sapwood	15	13.051 ± 0.888	16.246 ± 0.767	0.803 ± 0.032
Heartwood	24	12.849 ± 0.485	15.960 ± 1.082	0.805 ± 0.028

teen sapwood samples was 0.803 ± 0.032, and for the heartwood it was 0.805 ± 0.026 with a pooled mean of 0.804 ± 0.028. The somewhat lower mean value of approximately 0.80 obtained in this study compared with the values cited previously may be at least partially accounted for by the fact that the initial desorption isotherm was used here rather than the cyclic desorption value used by other workers. This gives a higher desorption EMC than the cyclic desorption value for the same humidity of 76 percent (Figure 2.28). Another possible factor is the presence of volatile constituents in the green samples which are driven off during drying in the oven at $103°C$, thus giving higher moist weights in the never-dried samples compared with those previously dried and then allowed to adsorb moisture.

Stamm (1964) gives A/D ratios for white spruce ranging from 0.79 to 0.86 over the relative humidity range from 10 to 90 percent with a mean of 0.83 and, from the data of Kelsey (1957) for Klinki pine values ranging from 0.74 to 0.83 with a mean of 0.78. Stamm also points out that extractives have little ef-

fect on the A/D ratios and that the ratios for the various isolated wood components are quite similar to those for wood.

According to Weichert (1963), sorption hysteresis decreases with increasing
wood temperature and disappears at temperatures of 75° and 100°C for European spruce. Kelsey (1957) also shows a reduction in hysteresis between 25°
and 55°C for *Araucaria klinkii* of Australia.

In addition to the adsorption-desorption hysteresis discussed above another
history-dependent factor affects the EMC of wood at a given humidity. For example, the EMC in adsorption is higher when a dry sample is exposed to a given
humidity in one single step than when it is brought to the same humidity
through a series of adsorption steps. This is shown clearly in Figure 2.30, taken
from Tarkow (1960), who derived the data from Christensen and Kelsey (1959).
Undoubtedly this phenomenon is a significant factor in determining the magnitude of the hysteresis ratio A/D at any given humidity. For example, the ad-

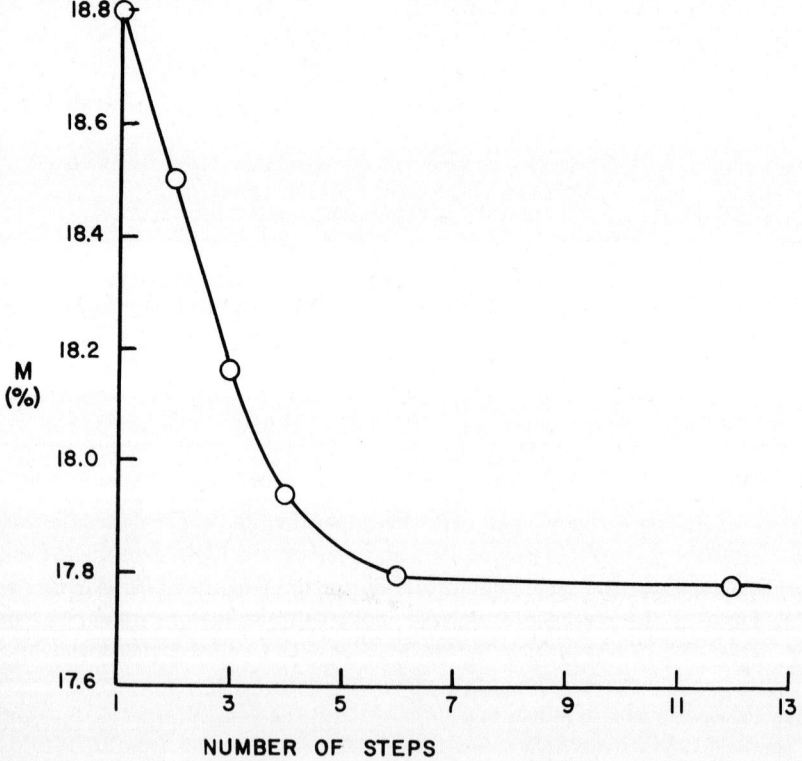

Fig. 2.30. Equilibrium moisture content M (%) as a function of number of intervening
steps during adsorption from the dry condition to a relative humidity H of 88 percent
(adapted from Tarkow 1960, using data of Christensen and Kelsey 1959).

sorption curve of Figure 2.28 which gives an EMC of approximately 18 percent at $h = 0.88$ shows five intervening adsorption steps from the dry condition to $h = 0.88$. The cyclic desorption curve gives an EMC about 3.5 percent higher, or 21.5 percent. The A/D hysteresis ratio therefore is 18.0/21.5 or 0.84. According to Figure 2.30 the adsorption EMC would be nearly one percent higher if only one step had been used in going from the dry condition to $h = 0.88$. Therefore the expected A/D ratio would be in the order of 19.0/21.5 or 0.88, compared with 0.84 obtained from the multistep data.

The findings of Christensen and Kelsey (1959) have been substantiated by Prichananda (1966), who showed (Figure 2.31) that the adsorption isotherm

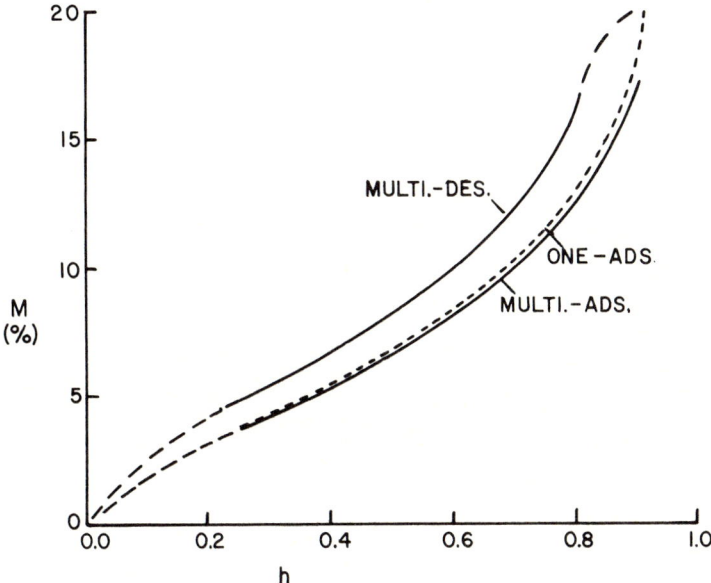

Fig. 2.31. Multistep desorption (MULTI DES), multistep adsorption (MULTI ADS), and one-step adsorption (ONE-ADS) isotherms for yellow birch (from Prichananda 1966).

for yellow birch (*Betula alleghaniensis* Britton) in air for single-step adsorption from the dry condition is higher than that obtained from the usual multistep process. From the desorption curve obtained by small steps beginning at 90 percent humidity, also shown in Figure 2.31, it is clear that the A/D hysteresis ratio is larger for the single-step than for the multistep adsorption curve. In the case of multistep adsorption the dry sample was exposed to successive increments of 10 percent after first being brought to equilibrium with 30 percent humidity in a single step.

The sorption isotherms for wood are also dependent on the temperature to

which the wood has previously been exposed. This effect will be discussed under the effect of temperature in later paragraphs.

Effect of Wood Species and Extractives

It is well known that the sorption isotherms vary from one kind of wood to
another. This may be the result of a number of factors such as differences
among woods with respect to the proportions of the major wood constituents
such as cellulose, hemicellulose, and lignin in different woods. For example,
Figure 2.32, adapted from Christensen and Kelsey (1959), shows adsorption

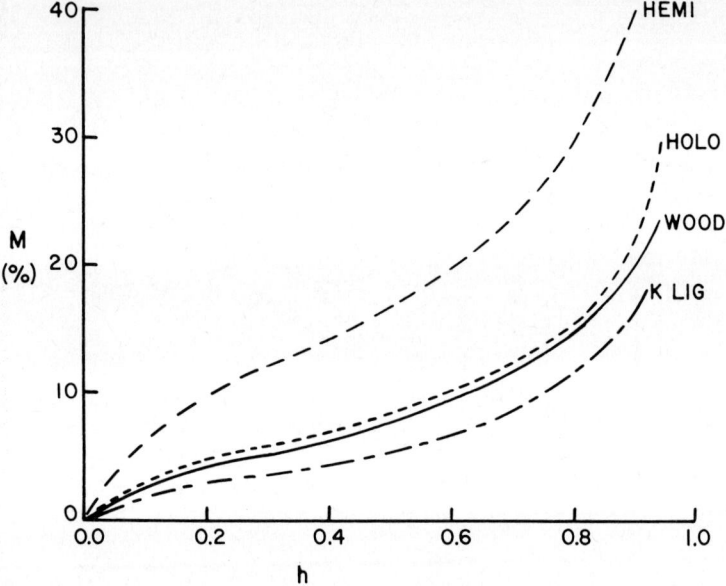

Fig. 2.32. Adsorption isotherms for wood hemicellulose (HEMI), holocellulose (HOLO),
Klason lignin (K LIG), and gross wood (WOOD) at 25°C (adapted from Christensen and
Kelsey 1959).

curves for representative holocellulose, hemicellulose, and lignin of *Eucalyptus
regnans* as well as that for the gross wood. They calculated that cellulose contributes 47 percent, the hemicellulose 37 percent, and lignin 16 percent of the
total water-sorption capacity of this wood. It should be remembered, however,
that the adsorption curves for the constituents may be different in the gross
wood than in the chemically isolated constituents because of the effect of ex-
tracting the constituents and also because of the interrelationships among the
constituents in the cell wall.

The differences among species of wood grown in the temperate regions of the world with respect to sorption behavior is not generally very great. There are exceptions, however, particularly in woods which have a high extractive content, such as redwood. Spalt (1958) calculated the moisture content M_f of the fully swollen cell wall to range from 30.3 percent for white spruce to 22.9 percent for redwood among eight United States softwoods, and from 33.6 percent for *Ceiba pentandra* to 21.8 percent for *Licaria cayennensis* among eight hardwoods, mostly tropical. The values of M_f were calculated by fitting the Hailwood and Horrobin equation, given in Chapter 5, to desorption curves for each wood and using the equation thus obtained to evaluate the EMC at 100 percent humidity. Presumably the variation among species with respect to the hygroscopicity of sapwood is considerably less.

Table 2.6, from Wangaard and Granados (1967), shows the extractive content

Table 2.6. Increase in Total Sorption M_f^* Following Removal of Extractives
(from Wangaard and Granados 1967)

	Extractive Content (%)	Adsorption (%)		Desorption (%)	
		$(M_f)_{ads}$	$\Delta(M_f)_{ads}$	$(M_f)_{des}$	$\Delta(M_f)_{des}$
Jacaranda copaia	5.93	25.92	3.38	32.78	2.12
Couratari pulchra	2.88	23.71	3.20	28.16	2.22
Minquartia guianensis	10.25	18.63	7.01	27.33	3.44
Tabebuia donnell-smithii	5.53	23.71	4.00	26.80	8.43
Pterocarpus vernalis	8.86	24.47	4.91	26.64	6.19
Goupia glabra	16.03	18.03	8.74	25.06	7.98
Hibiscus elatus	7.67	24.01	2.99	24.95	10.15
Swietenia macrophylla	14.90	23.01	5.89	24.28	13.76
Chlorphora tinctoria	17.05	15.42	11.17	20.52	12.31

*The value of M_f was obtained in each case by fitting the adsorption and desorption isotherms to the Hailwood and Horrobin equation (see Chapter 5) and using this equation to evaluate the EMC at 100 percent humidity.

and the increase in total sorption (equivalent to the fiber-saturation point M_f) of nine tropical woods following consecutive extractions in 2:1 benzene-alcohol (95 percent), 95 percent alcohol, and distilled water for ten to twenty days using a soxhlet apparatus. The table indicates that the increase in sorption ΔM_f decreases with increasing total moisture M_f in the wood at saturation in the unextracted condition. Based on this relationship, Wangaard and Granados calculated the following linear regression equations

$$\Delta(M_f)_{ads} = 27.58 - (M_f)_{ads} \tag{2.13}$$

and

$$\Delta(M_f)_{des} = 33.68 - (M_f)_{des} \tag{2.14}$$

The high extractive content and associated reduced hygroscopicity of certain woods such as redwood and the cedars is believed to be one factor which contributes to their good dimensional stability under exposure to cyclic humidity changes. This topic is discussed in more detail under shrinking and swelling in Chapter 3.

Effects of Temperature

Temperature affects the sorption isotherm of wood in two ways—the immediate effect of temperature and the effect of temperature history.

The immediate effect of temperature, as shown in Figure 2.33, is to reduce

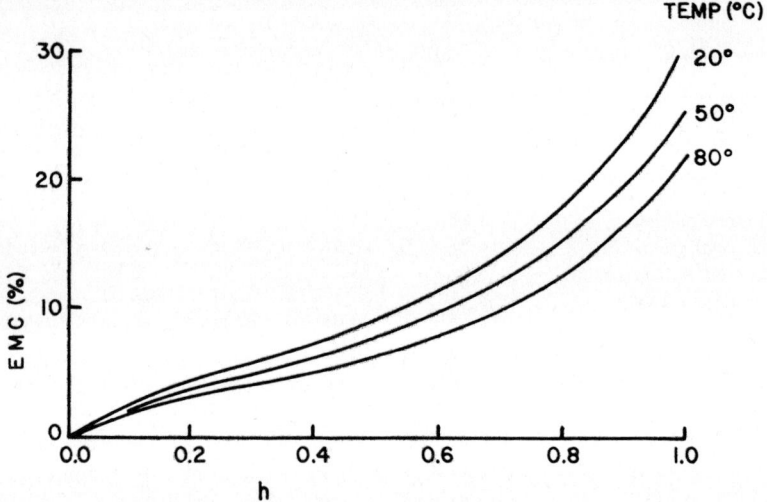

Fig. 2.33. Sorption isotherms at three different temperatures showing EMC (%) against relative vapor pressure h.

the hygroscopicity of wood at a given relative humidity. This effect is related to the thermodynamics of sorption, which is discussed in more detail in Chapter 4. It can be seen from Figure 2.33 that the fiber-saturation point M_f obtained by extrapolating the sorption isotherm to 100 percent relative humidity decreases with increasing temperature. According to Stamm and Loughborough (1935), this decrease is approximately 0.1 percent per °C increase in temperature between 25° and 100°C for Sitka spruce. Kollmann (1959) presents curves of Lykow which shows a decrease in M_f from 31 percent at 20°C to 23 percent at 80°C, which is (31 - 23)/(80 - 20) or 0.13 percent per °C. Weichert (1963) shows a decrease in M_f from 30 percent at 25°C to 21 percent at 100°C or 0.12

percent per °C. The opposite effect is expected at temperatures below 0°C for reasons discussed in Chapter 4 in connection with coldness shrinkage.

Exposure of dry wood to high temperatures for various lengths of time causes a permanent decrease in hygroscopicity, accompanied by a weight loss below the normal ovendry weight, resulting from a loss in water of constitution. According to Stamm (1964), most of the decrease in hygroscopicity is probably due to the decomposition of the hemicelluloses, which are more hygroscopic than the other primary wood constituents (Figure 2.32) and also degrade more rapidly at elevated temperatures than do cellulose or lignin (Beall 1968). Figure 2.34 shows

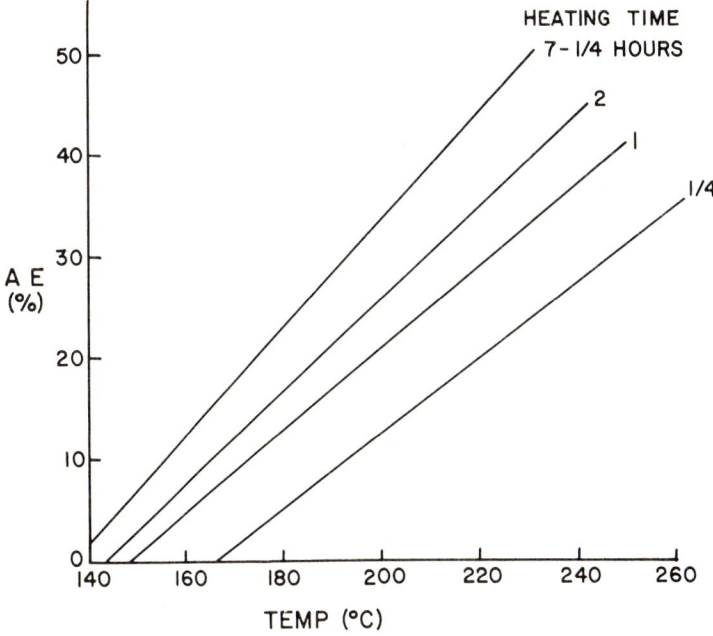

Fig. 2.34. Antishrink efficiency, AE (%) against heating temperature (°C) for several heating times (hours) (adapted from Stamm, Burr, and Kline 1955).

the reduction in hygroscopicity expressed in terms of anti-shrink efficiency between the humidity limits of 30 and 90 percent as given by Stamm, Burr, and Kline (1955) for Western white pine heated under molten metal for various combinations of time and temperature. Anti-shrink efficiency is defined in this case as the difference between the relative weight changes of an untreated specimen and of a heat-treated specimen, divided by that of the untreated specimen, expressed as a percentage. Anti-shrink efficiency is defined basically in terms of relative shrinkages but has been found to be essentially the same as the values obtained from the definition given above.

It would appear that heating dry wood should be a good method for reducing the hygroscopicity and resultant shrinking or swelling with humidity changes. Unfortunately, however, there are permanent losses in weight and in certain desirable physical and mechanical properties which mitigate against the use of this technique. A detailed discourse on methods of stabilizing wood against changes of atmospheric moisture is outside the scope of this text but can be found in Stamm (1964).

Wood which has been kiln dried usually has a lower EMC than air-dried wood. Furthermore, the EMC is generally lower the higher the kiln temperature used. For example, Salamon (1966) found for one series of experiments that the average EMC for western hemlock at 72°F and 60 percent RH was 9.65 percent for air-dried samples, 9.11 percent for samples dried at temperatures from 160° to 185°F, and 8.39 percent for samples dried at 225°F dry-bulb temperature.

Miscellaneous Factors

Other factors in addition to those discussed above affect the hygroscopicity of wood. Among the more interesting is the effect of stress, which was first explained by Barkas (1949). This effect predicts a lowering of the EMC of wood when it is subjected to a compressive stress and the reverse effect when a tensile stress is applied. It is most pronounced when a given stress is applied in the direction in which maximum hygroscopic swelling takes place, that is, in the tangential direction for normal wood. Furthermore, it also causes a reduced moisture sorption in wood which is restrained from swelling when it is exposed to a higher humidity. For example, Bello (1968) found that the mean EMC of five United States hardwoods which were restrained from transverse swelling by means of steel rings, and allowed to adsorb moisture, from the dry condition to equilibrium with 87 percent humidity, was 16.69 percent. This was 1.44 percent lower than the EMC of the same woods which were unrestrained from swelling. This factor is discussed in more detail in Chapter 4.

Some studies have been made on the effect of nuclear radiation on the hygroscopicity of wood. Most of this effort has been concentrated on the effect of gamma radiation. Figure 2.35, taken from Paton and Hearmon (1957), indicates that a dosage rate of 10^8 rads of gamma radiation consistently lowers the EMC over the entire sorption range. This effect is smaller for smaller dosages of radiation.

Moisture Content of Wood in Use

When wood is used in its final form as a part of furniture or as a building material, for example, it is subsequently subjected to environmental conditions

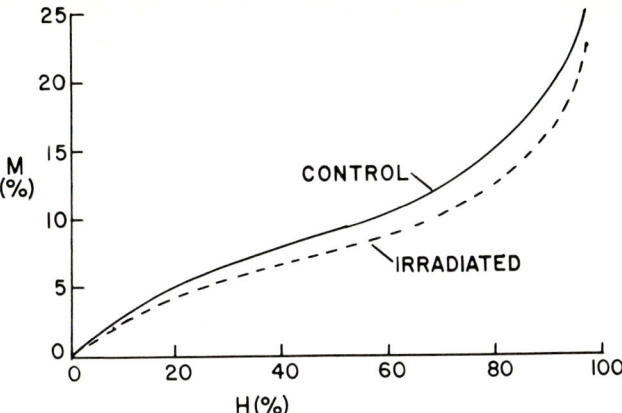

Fig. 2.35. Effect of 10^8 rads of gamma radiation on the sorption isotherm of wood (adapted from Paton and Hearmon 1957).

which generally fluctuate cyclically, such as a diurnal change related to the daily cycle of temperature and humidity or an annual change associated with the seasons of the year. Therefore, the EMC conditions also fluctuate periodically as well as in the more random fashion caused by such factors as rainstorms.

It is clear from Figure 2.33 that the EMC of wood in use is essentially dependent on the temperature and humidity of the atmosphere to which it is exposed, the latter being the most important single factor over the normal range of atmospheric conditions. Relative humidity varies markedly from one geographical location to another throughout the world and even within relatively short distances where mountain ranges, for example, may be an important factor.

In the temperate climates where buildings are heated in winter there are large variations in relative humidity indoors simply because of outdoor temperature variations. For example, Figure 2.36, based on Hoadley (1967), shows the variation throughout the year of both the outdoor and indoor relative humidities and also of the EMC of wood in both locations. The indoor humidities were calculated from the monthly average outdoor values of temperature and relative humidities at Amherst, Massachusetts, assuming a constant indoor temperature of $70°F$ ($21°C$) and a vapor pressure indoors p' equal to the outdoor vapor pressure p''. The indoor relative vapor pressure (p'/p_0') is therefore calculated from the outdoor relative vapor pressure (p''/p_0'') by the equation

$$p'/p_0' = p''/p_0' = (p''/p_0'')(p_0''/p_0') \tag{2.15}$$

where p_0' and p_0'' are the indoor and outdoor saturated vapor pressures. Since p_0' is larger than p_0'' in the winter months the indoor relative vapor (p'/p_0') is lower than that outdoors (p''/p_0'') even when the actual vapor pressure p' and p'' are equal.

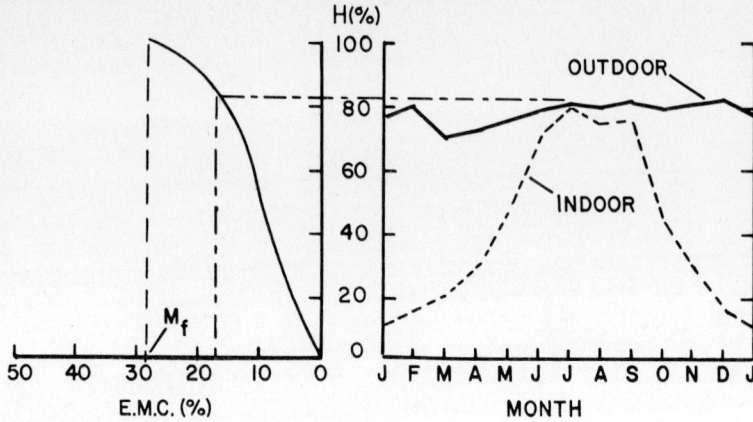

Fig. 2.36. Annual variation of outdoor and indoor relative humidity H (%) with month of year and of corresponding equilibrium moisture content, EMC (%), of wood at Amherst, Massachusetts (adapted from Hoadley 1967).

Based on Figure 2.36 it appears that the EMC outdoors in the Massachusetts climate is essentially constant at approximately 16 percent but that the EMC indoors may fluctuate from a low of 2 percent in midwinter to the outdoor equilibrium of 16 percent in midsummer. This does not necessarily mean that particular wood samples such as the components of wooden furniture kept indoors in Amherst will vary between the two extremes given above. This situation is affected by several factors, including the long times required for moisture to diffuse in and out of thick lumber and the retarding effect of protective coatings, as Hoadley points out.

Another factor which operates to reduce the seasonal moisture variations is hysteresis in the sorption isotherm. This effect, already discussed previously, is maximum when the sorption cycle occurs over the entire range of h from 0 to 1.0. In this case the EMC during adsorption is about 0.85 of the value during desorption over most of the sorption range.

When wood is exposed to smaller humidity cycles, the hysteresis effect is less pronounced but still has the effect of reducing the change in EMC associated with a given change in p/p_0. For example, in Figure 2.28 the cyclic change in EMC between p/p_0 limits of 0.80 to 0.20 is 16–6 percent rather than 19–6 percent as it would be for desorption only, or 16–5 percent as it would be for adsorption only. This represents a cyclic change in EMC of only 10 percent rather than 13 percent for the desorption curve and 11 percent for the adsorption curve, or an average change of 12.0 percent in EMC for an intermediate curve. A similar reduction in moisture change for a given humidity change is also anticipated because of the gradual nature of the cyclic changes, according to the findings of Christensen and Kelsey (1959) as discussed above.

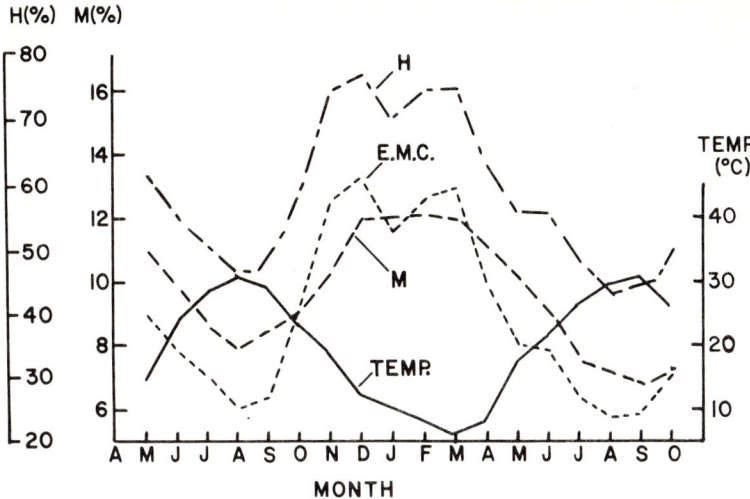

Fig. 2.37. Annual cyclic variation in relative humidity *H* (%), temperature (°C), equilibrium moisture content EMC (%), and measured mean wood moisture content *M* (%) in a sheltered outdoor location in Thessalonika, Greece (adapted from Tsoumis 1960).

A study was made by Tsoumis (1960) of seasonal moisture-content variations in previously air-dried wood test samples of four different European species—beech, oak, chestnut, and pine—exposed to the *outdoor* atmosphere under shelter in Thessalonika, Greece. His results, summarized in Figure 2.37, show a similar cyclic annual moisture-content variation which is more or less opposite to the seasonal variations shown in Figure 2.36. In other words, the EMC is maximum in winter and minimum in summer because the relative humidity in Thessalonika follows the same seasonal cycle, high in the winter and low in the summer. Figure 2.37 shows the seasonal temperature variations and also the calculated EMC, the latter presumably based on the sorption isotherm. It is interesting to note that observed moisture-content variations are less than the calculated EMC variations, as anticipated.

It is clear that the moisture content to which wood should be dried prior to use depends on the location where it will be exposed during use. It should be dried to equilibrium with the mean conditions to which it is to be exposed in use. These conditions vary geographically, even within the same country. For example, Table 2.7, taken from Peck (1950), shows the estimated average moisture content of interior woodwork in thirteen widely separated cities in the United States during January and July, months which may be taken to show the seasonal extremes. Table 2.8, also taken from Peck (1950), shows the recommended moisture content of lumber intended for either interior or exterior use in the three principal climatic regions of the United States.

Table 2.7. Estimated Average Moisture Content of the
Principal Interior Woodwork in 13 Widely Separated
Cities During January and July (from Peck 1950)

City	Moisture Content of Interior Woodwork	
	July (%)	January (%)
Atlanta, Georgia	11.5	8.5
Albuquerque, New Mexico	6.0	7.0
Boston, Massachusetts	13.0	7.0
Dallas, Texas	9.0	9.0
Duluth, Minnesota	10.5	5.0
Madison, Wisconsin	10.0	6.0
New Orleans, Louisiana	13.5	12.5
New York, New York	12.5	7.0
Portland, Oregon	9.5	9.0
Salt Lake City, Utah	4.0	7.0
San Francisco, California	10.5	10.5
Seattle, Washington	11.0	8.5
Washington, D.C.	11.0	8.0

Table 2.8. Recommendations for Initial Moisture Content of Lumber
for Dwellings (from Peck 1950)

Use of Lumber	Moisture content (percent of weight of ovendry wood)					
	Dry Southwestern States		Damp Southern Coastal States		Remainder of the U.S.	
	Average	Tolerance	Average	Tolerance	Average	Tolerance
Interior finishing woodwork and flooring	6	4–9	11	8–13	8	5–10
Sheathing, framing,* siding, and exterior trim	9	7–12	12	9–14	12	9–14

*Framing lumber of higher moisture content is commonly used in ordinary construction because material of the moisture content specified may not be available except on special order.

3. Hygroscopic Shrinking and Swelling of Wood

One of the most important practical problems which arises during the use of wood is the hygroscopic shrinking and swelling of wood which occurs as a result of moisture changes. In the living tree this is no problem, as Tiemann (1944) points out, because the cell walls of green wood are always in the fully swollen condition; that is the green moisture content is higher than the fiber-saturation point. Therefore there is no hygroscopic wood shrinkage in the living tree except for the small amount which occurs as a result of changes in fiber-saturation point with temperature as is discussed in Chapter 2.

When green wood dries, however, shrinkage occurs, the magnitude of which depends on a number of factors. These include extent of moisture loss, structural direction (radial, tangential, or longitudinal), wood density or specific gravity, temperature, degree of drying stress caused by moisture gradients, and other factors. Several of the more important of these factors will be discussed in the pages to follow, after first considering the hygroscopic shrinking and swelling of the cell wall itself.

Volumetric Shrinkage and Swelling of the Cell Wall

Density of the Dry Cell Wall

Before proceeding with a discussion of the actual swelling and shrinking of the cell wall of wood it is necessary to consider first the density ρ_0' of the dry cell wall. This has a value of approximately 1.5 g/cc when measured by pycnometric or volume-displacement methods. The exact value obtained depends upon the medium used for measuring the volume. When a polar swelling liquid such as water is used as the displacement medium the apparent specific volume V_{0w}' of the dry cell wall is lower than the value V_{0n}' obtained when a nonswelling medium such as toluene is used, as shown in Table 3.1. For example, the four woods measured by Wilfong (1966) shown in Table 3.1 give a mean apparent speciifc volume V_{0w}' of 0.656 cc/g using water displacement and a V_{0n}' of 0.689 using toluene displacement. These correspond to densities of 1.525 and 1.451 g/cc, respectively.

Table 3.1. Apparent Density ρ_0' and Specific Volume V_0' Obtained for the
Dry Wood Cell Wall Using Various Displacement Media

Kind of Wood	Displacement Liquid	Apparent Density of Dry Cell Wall ρ_0' (g/cc)	Apparent Specific Volume (cc/g)	Author
Alaska cedar	Water	1.548	0.646	Stamm (1964)
Alaska cedar	Ethanol	1.537	0.651	
Alaska cedar	Benzene	1.476	0.678	
Alaska cedar	Mineral oil	1.460	0.685	
Abies grandis	Water	1.52 ± 0.03	0.658	Petty (1971)
Abies grandis	Toluene	1.44 ± 0.03	0.694	
Engel. spruce (unextracted)	Water	1.524	0.656	Wilfong (1966)
Engel. spruce (unextracted)	Toluene	1.448	0.691	
Engel. spruce (unextracted)	Helium	1.449	0.690	
Doug.-fir (Htwd) (unextracted)	Water	1.540	0.649	Wilfong (1966)
Doug.-fir (Htwd) (unextracted)	Toluene	1.453	0.688	
Doug.-fir (Htwd) (unextracted)	Helium	1.464	0.683	
Redwood Htwd. (unextracted)	Water	1.505	0.664	Wilfong (1966)
Redwood Htwd. (unextracted)	Toluene	1.437	0.696	
Redwood Htwd. (extracted)	Water	1.532	0.653	Wilfong (1966)
Redwood Htwd. (extracted)	Toluene	1.467	0.682	

Figure 3.1 illustrates the difference $V_{0n}' - V_{0w}'$ between the specific volumes of the cell wall obtained using the two types of displacement media. Stamm (1964) gives two reasons to account for this difference. One is the greater penetrability of the water molecule into microvoid spaces in the cell wall and the other is the compaction or apparent reduced volume of the sorbed water compared with free liquid water. Both of these effects tend to reduce the apparent specific volume of the wood cell wall when water is used as the displacement medium. The latter effect is similar to the phenomenon observed when two miscible solvents, such as ethanol and water, are mixed. If 100 cc of each of these are mixed, the resulting solution has a volume of only 192 cc, not 200 cc, at 15.56°C, which indicates that the volume of the solution is 4 percent lower than that of the two components taken separately.

Based on the figures given above for the ethanol-water mixture, if one assumes

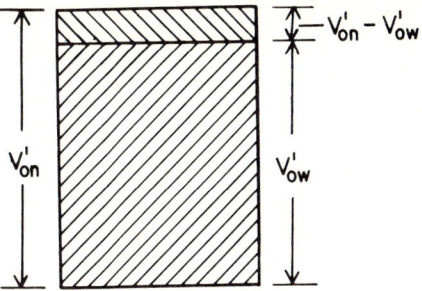

Fig. 3.1. Difference in the specific volume of the dry cell wall using water displacement V'_{ow} (cc/g) and a nonswelling medium V'_{on} (cc/g).

that all of the 8 cc loss of volume upon mixing is due to an apparent compression in the water, this can be calculated to be 8/100 or 0.08 cc/g. Stamm (1964) gives such a calculation for data obtained on spruce, assuming that the difference in the specific volumes of 0.685 and 0.653 cc/g, obtained using water and helium as displacement media (see Figure 3.2), was entirely due to apparent compaction of the water. For this calculation it was necessary to know the saturation moisture content m_f of the cell wall which he took as 0.31 g/g. This amount of free water has a normal volume of 0.31 cc, but the apparent decrease in volume of the water in the cell wall per gram of dry wood was 0.685 - 0.653, or 0.032 cc. This gives an apparent water volume of 0.310 - 0.032 or 0.278 cc, and an apparent density therefore of 0.310/0.278, or 1.115 g/cc, for the water in the cell wall. Figure 3.2 illustrates the principles involved.

Stamm (1964) also gives a small difference in the specific volumes of the dry cell wall of spruce using benzene and helium as displacement media, the values

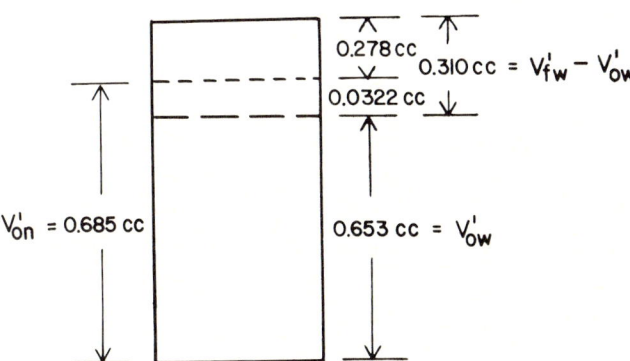

Fig. 3.2. Apparent compaction of water in the cell wall based on differences in shrinkage from the fully swollen specific volume V'_{fw} (cc/g) to the dry volume based on water displacement V'_{ow} (cc/g) and on displacement in a nonswelling medium V'_{on} (cc/g).

being 0.693 and 0.685 cc/g, respectively. Later work by Wilfong (1966), based on improved techniques, showed no consistent differences between the specific volumes for the dry cell wall using helium and toluene.

Weatherwax and Tarkow (1968) have attempted to separate the two factors which account for the difference in specific volume obtained by water and by nonswelling displacement media, rather than attributing it all to the apparent change in specific volume of the water. They measured the specific volume V_0' of the dry cell wall of Sitka spruce using the nonswelling displacement medium silicone oil. This is equivalent to the value $V_{0n}'(0.6825 \text{ cc/g})$ obtained using toluene, for example. It includes the inaccessible microvoids in the cell wall shown in Figure 3.3. They also measured $V_{0w}'(0.6470 \text{ cc/g})$, the specific volume using water displacement. The difference between these (0.6825 - 0.6470) is 0.0355 cc/g which they reasoned consisted of two parts—one the apparent adsorption compression* of the water, and the other microvoids in the cell wall which are penetrated by the water but not by the nonswelling silicone oil.

In order to measure the volume of the microvoids Weatherwax and Tarkow first swelled the wood sections with water. This they displaced with nonswelling hexane through an intermediate solvent exchange using ethanol, which is soluble in both water and hexane and also swells wood to a similar extent as water. In this way the hexane-saturated wood was in a swollen condition allowing the nonpolar hexane to penetrate the microvoids in the cell wall. The specific volume thus obtained (see Table 3.2) was 0.6522 cc/g which is 0.0303 cc/g smaller than the value (0.6825) obtained using silicone oil.

These experiments were interpreted to mean that the total difference (0.0355 cc/g) in specific volume between the water-displacement (0.6470 cc/g) and silicone-oil displacement (0.6522) values consisted of 0.0303 cc/g void volume, and 0.0052 cc/g (0.0355 - 0.0303) apparent compression of the sorbed water. Thus the fractional void space in the cell wall is 0.0444 (obtained from 0.0303/

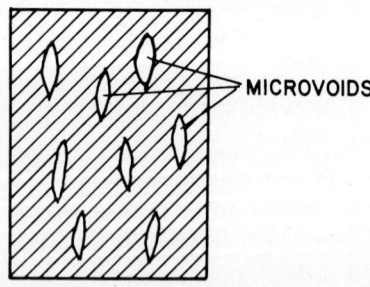

Fig. 3.3. Schematic diagram of cell wall showing microvoids inaccessible to nonswelling displacement media.

---

*This factor is described by Goring (1966) as "perturbation of the water structure" in the cell wall compared with that of ordinary liquid water.

Table 3.2. Apparent Dry Density ρ_0', Specific Volume V_0', and Reductions
in Specific Volume Based on the Silicone Oil Value
(adapted from Weatherwax and Tarkow 1968)

Displacement Medium	Apparent Density ρ_0' (g/cc)	Specific Volume V_0' (cc/g)	Apparent Reduction in V_0' Based on Silicone Oil Value (cc/g)	Fractional Reduction in V_0' Based on Silicone Oil Value (cc/g)
Water	1.5457	0.6470	0.0355	(0.0355/0.6825) = 0.0520
Hexane*	1.5333	0.6522	0.0303	(0.0303/0.6825) = 0.0444
Silicone Oil	1.4650	0.6825	0.0000	—

*The values obtained using hexane are after solvent displacement from water-swollen wood through ethanol to hexane, which has therefore penetrated into the cell-wall microvoids.

0.6825), and the fractional loss in cell-wall volume based on apparent compression of the sorbed water is 0.0076 (0.0052/0.6825). The sum of the two fractional components (0.0444 + 0.0076) is 0.0520. Weatherwax and Tarkow concluded that about 85 percent (0.0444/0.0520) of the difference in specific volume of the dry cell wall obtained by water- and by nonswelling-liquid displacement is caused by the lower accessibility of nonswelling liquids to microvoids in the cell wall. Therefore only 15 percent (0.0076/0.0520) is caused by the apparent compression of the sorbed water.

In order to calculate the apparent density ρ_w' of the water in the cell wall it is necessary to know the fiber-saturation point M_f. If this is taken as 31 percent, then the apparent density of the sorbed water is $[0.31/(0.3100 - 0.0052)]$ = (0.31/0.3048) = 1.017 g/cc. If it is assumed that there are no microvoids in the cell wall and that all of the volume difference obtained between water displacement and silicone-oil displacement is caused by apparent compression of the sorbed water, the apparent density of the sorbed water is $[0.31/(0.3100 - 0.0355)]$ = 1.129 g/cc. The latter value is close to the value of 1.115 g/cc calculated from Stamm's data, but the value of 1.017 g/cc appears to have a more favorable theoretical basis since it considers the microvoid volume.

In recent years there has been a renewed interest in the density of the cell wall of wood, and several new techniques as well as the classical pycnometric techniques have been used. Stamm (1967) and Wangaard (1969) have reviewed some of these developments. Wangaard has proposed that the dry density of "wood substance" be taken to mean the value for ρ_0' obtained by pycnometric techniques using water as the displacement medium (corrected for apparent compression), and that the lower value obtained using nonswelling liquids such as benzene or toluene be designated as the density of the "cell wall" of wood. There is merit in this suggestion since cell-wall density measurements made by optical methods, discussed below, include whatever microvoids may exist in the cell wall since these are not visible with the optical microscope. Wangaard also rejects the contentions of some recent workers who have proposed that, based on optical measurements of cell wall densities, there are substantial microvoid volumes in the dry cell wall of wood. However, he proposes that the differences in specific volumes of the cell wall using water and toluene displacement media may be considered to be due in part to the apparent "compression" of the water and in part to the cell-wall microvoid volume, in agreement with the findings of Weatherwax and Tarkow (1968), as discussed above.

One of the techniques for measuring cell-wall density which has been used frequently in recent years is the optical method. This involves measuring the relative fractions of cell-wall and void volumes in thin microtomed cross-sections of dry or moist wood, using special microscopic methods. In the case of dry wood this information is then combined with measurements of the dry density ρ_0 of the gross wood sample to calculate the dry cell-wall density ρ_0', using the equation

$$\rho_0' = \rho_0(V_0/V_0') \tag{3.1}$$

where V_0 and V_0' are the specific dry volumes of the gross wood and of the cell wall. For a thin microtome section of uniform thickness V_0'/V_0 is the fraction of the cross-section occupied by the cell wall.

Kellogg and Wangaard (1969) used such a microscopic technique to calculate ρ_0' for eighteen different woods. Upon comparing these values with those obtained by pycnometric methods on the same woods they noted that the apparent mean specific volume obtained optically for seven species was 0.777 cc/g. After adjusting the individual values for differences in cell-wall and cell-lumen shrinkages in microtomed sections compared with larger wood blocks, which were used to obtain ρ_0 in equation (3.1), they found that the adjusted mean value obtained optically was 0.704 cc/g ($\rho_0' = 1.420$ g/cc). This was higher than the value of 0.679 cc/g ($\rho_0' = 1.473$ g/cc) obtained pycnometrically using toluene as the displacement medium. They attributed the difference to various uncontrollable factors such as cell-wall ruptures in the microtomed sections, which tended to give an overestimate of the specific volumes using this technique. Another possible explanation for the lower values of cell-wall density determined by the optical method compared with the pycnometric method may be related to a kind of parallax effect in the optical method. According to Petty (1971), there is a tendency to overestimate the fraction of cell-wall volume due to a "shadow effect" and therefore to underestimate the cell-wall density. This effect becomes more important as the thickness of the microtomed section increases.

Maximum Swelling and Shrinking of the Cell Wall

Before discussing maximum swelling and shrinking of the cell wall between the dry and fully swollen condition, we must first define the term "specific gravity,"

G, as it is applied to wood in the United States and other North American countries. For most materials, specific gravity is defined as the ratio of the density of the material to that of water at some specified temperature, usually 4°C, or as the ratio of the mass or weight of the material to that of an equal volume of water. This same definition applies for dry wood, but for wood containing water a peculiar variation of this definition is used, in that the dry weight w_0 and swollen volume v_m basis is used in defining the specific gravity G_m at the wood moisture content M. In other words, the specific gravity for wood is a hybrid type of density index and perhaps should be called "density index" as Panshin and de Zeeuw (1970, p. 209) have suggested. Because of widespread traditional usage and acceptance, however, the term "specific gravity" will be used here to mean the ratio of the dry weight w_0 to the weight of water displaced by the wood at its volume v_m at moisture content M, or

$$G_m = (w_0/V_m)/\rho_w \qquad (3.2)$$

where ρ_w is the density of water at the temperature specified, usually room temperature.

When the dry cell wall is immersed in water it swells in proportion to the total apparent volume v'_w of water taken up, less the water which fills the void spaces, as is shown in Figure 3.3. Using the specific volume V'_{0w} of the dry cell wall based on *water displacement* as a basis for calculating the total volumetric swelling S'_f of the cell wall from the dry condition to fiber saturation M_f we obtain

$$S'_f = 100(v'_{wf})/(V'_{0w}w'_0) \qquad (3.3)$$

where v'_{wf} is the volume of the water in the cell wall at saturation and w'_0 is the dry mass of the cell wall. This reduces to

$$S'_f = 100(w'_{wf}/\rho'_{wf})/(V'_{0w}w'_0) = M_f G'_{0w} \qquad (3.4)$$

since v'_{wf} is equal to the weight w'_{wf} of water in the cell wall divided by its density ρ'_{wf}; M_f, the fiber-saturation point, is equal to $100 \, w'_{wf}/w'_0$; and the specific gravity G'_{0w} of the dry cell wall based on water displacement is equal to the reciprocal of $\rho'_{wf}V'_{0w}$, since ρ'_{wf} is assumed to be equal to the density ρ_w of free water when G'_0 is based on water displacement.

The specific gravity G'_f based on dry weight w'_0 and swollen volume v'_f of the cell wall is defined as

$$G'_f = (1/\rho_w) \, [w'_0/(v'_{0w} + v'_w)] \qquad (3.5)$$

where the sum $v'_{0w} + v'_w$ is equal to v'_f. Equation (3.5) can be reduced to

$$G'_f = G'_{0w}/(1 + m_f G'_{0w}) \qquad (3.6)$$

where $m_f = M_f/100$, or, using equation (3.4) it becomes

$$G'_f = G'_{0w}/[1 + (S'_f/100)] \tag{3.7}$$

in terms of the percent swelling S'_f.

The density ρ'_f of the cell wall at full saturation can be defined as

$$\rho'_f = w'_f/(v'_{0w} + v'_w) = G'_f \rho_w (1 + m_f) \tag{3.8}$$

or

$$\rho'_f = \rho'_{0w} \{(1 + m_f)/[1 + (S'_f/100)]\} \tag{3.9}$$

where w'_f is the weight or mass of the fully swollen cell wall. Equations (3.7) and (3.9) can be written in terms of the ratios of the dry and fully swollen specific gravities and densities, as follows

$$G'_f/G'_{0w} = 1/[1 + (S'_f/100)] = 1/(1 + m_f G'_{0w}) \tag{3.10}$$

and

$$\rho'_f/\rho'_{0w} = (1 + m_f)/(1 + S'_f/100) = (1 + m_f)(G'_f/G'_{0w}). \tag{3.11}$$

It is clear from equations (3.10) and (3.11) that G'_f/G'_{0w} is less than unity. For example, using the water-displacement data of Weatherwax and Tarkow (1968) from Table 3.2 ($G'_{0w} = 1.5467$) and assuming $m_f = 0.30$, the ratio becomes

$$G'_f/G'_{0w} = 1/[1 + (0.30)(1.5457)] = 0.683.$$

However, the ratio ρ'_f/ρ'_{0w} is also less than unity, since

$$\rho'_f/\rho'_{0w} = (1 + 0.30)(0.683) = 0.888.$$

The density of the cell wall decreases with increasing moisture content because the increase in weight or mass with increasing moisture content is less than the increase in volume. This does not usually occur in the case of the gross wood, however, as will be shown later in this chapter. The specific gravity always decreases when the cell wall absorbs moisture because the dry weight is constant and the volume always increases with moisture. This is also the case for the gross wood.

An equation for percent shrinkage s'_f of the fully swollen cell wall to the dry condition can also be obtained. It is

$$s'_f = M_f G'_f. \tag{3.12}$$

The relationship of percent shrinkage to percent swelling to the dry-volume and fully swollen-volume specific gravities is obtained by combining equations (3.4) and (3.12). Thus

$$s'_f/S'_f = G'_f/G'_{0w} \tag{3.13}$$

which is equal to 0.683 based on the example in the preceding paragraph.

It is clear from equation (3.6) that there is a relationship between the dry-volume and swollen-volume specific gravities and the saturation moisture content m_f. Equation (3.6) can be rearranged to evaluate m_f in terms of G_{0w} and G'_f. Thus,

$$m_f = (1/G'_f) - (1/G'_{0w}) = \rho_w (V'_f - V'_{0w}).\qquad(3.14)$$

If the two specific gravities could be measured independently for a given wood it should be possible to calculate the magnitude of m_f. This has been done by Kellogg and Wangaard (1969) for eighteen woods native to the United States. They measured G'_{0w} pycnometrically using water displacement as required, and measured G'_f for fully water-swollen microtome sections by *optical* methods. Substituting the mean values for the eighteen species into equation (3.14) gives as the fiber-saturation point, $m_f = (1.0036 - 0.6503) = 0.3533$, or $M_f = 35.33$ percent.

Kellogg and Wangaard actually used a variation of equation (3.14) which gave them slightly higher values than are obtained here. They assumed that the value of m_f as obtained by equation (3.12) was actually the apparent volume of water per gram of dry wood. Therefore, they multiplied this value by the apparent density of the sorbed water which they took from the work of Weatherwax and Tarkow (1968) to be equal to $m_f/(m_f - 0.0052)$. This was only a small correction in the order of 1.015 g/cc on the average for their samples, ranging from 1.009 to 1.022 g/cc. For example, if m_f as calculated above is 0.3533 the corrected value is $(1.015)(0.3533)$ or 0.3586.

They found a wide range in M_f, from 27.6 to 55.1 percent. Furthermore, M_f was greater for woods of lower density as Figure 3.4 shows, indicating that the fiber-saturation point decreases with increasing wood density. This is in agreement with the findings of Feist and Tarkow (1967), based on polymer exclusion in wood, and also those of Vorreiter (1963), which were based on data obtained by several workers who gave measured fiber-saturation points and also densities for a variety of different woods.

Petty (1971), however, has pointed out that there is a tendency to overestimate the specific volume of the cell wall by optical methods such as those used by Kellogg and Wangaard because of parallax problems at the edges of the cell walls. If compensation is made for this factor it would lower the values of M_f calculated by use of equation (3.14), since the term $1/G'_f$ which was measured optically would be reduced somewhat, other factors remaining constant. Furthermore, it is probable that this effect is more pronounced in the case of lower density woods since these would be expected to have thinner cell walls and therefore a larger ratio of lumen surface area to cell-wall volume. For this reason it is anticipated that the effect of wood density on fiber saturation would be less pronounced than is indicated by the calculations of Kellogg and Wangaard. Their published results, together with the curves of Feist and Tarkow (1967) and of Vorreiter (1963), are shown in Figure 3.4. According to A. J. Stamm (personal communication), the low values of the fiber-saturation point for woods of high

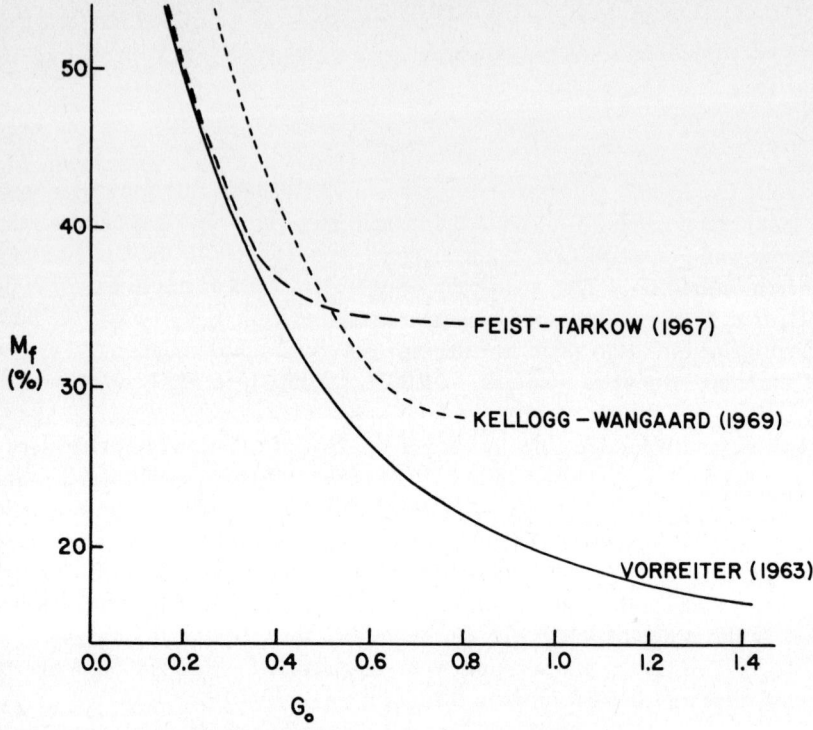

Fig. 3.4. Fiber-saturation point M_f (%) of wood as a function of dry-volume specific grav-ity according to Vorreiter (1963), Feist and Tarkow (1967), and Kellogg and Wangaard (1969).

specific gravity as given by the curve of Vorreiter (1963) in Figure 3.4 are at least partially caused by the bulking effects of the high extractive contents in these woods (see Chapter 2).

Maximum Volumetric Swelling and Shrinking of Gross Wood

When wood takes up moisture into the cell wall, the walls swell volumetrically in proportion to the volume of water absorbed. The volumetric swelling of the *gross* wood, that is the wood including air spaces, depends on the dimensional changes that occur in the air spaces or cell cavities when the cell wall swells. There are three possibilities which must be considered according to Tiemann (1944). These are shown schematically in Figure 3.5. First, the lumens or cavities may shrink so that all of the swelling, or at least part of it, takes place into the cell cavities (Figure 3.5b). Second, the lumens may remain constant in

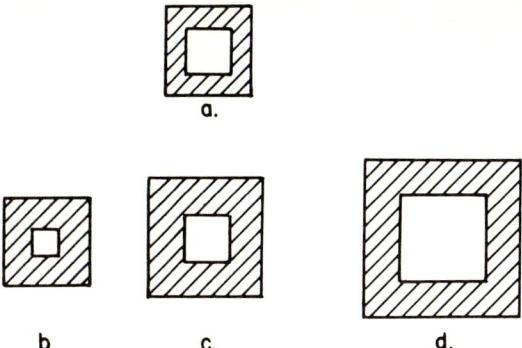

Fig. 3.5. Volumetric swelling of a single cell showing the cell, *a.* before swelling, *b.* with internal swelling into cell cavity only, *c.* with external swelling only, and *d.* with cell-cavity and external swelling.

size (Figure 3.5c). And third, the lumens may swell either in the same proportion as the cell wall or to a greater or lesser degree (Figure 3.5d).

Conceivably, all or part of the swelling may take place *into the cell cavity*, as shown in Figure 3.5b with *reduction in lumen volume*. The external swelling of the gross wood is then relatively small. If all of the swelling takes place into the cell cavities, no external dimensional changes take place at all. This extreme case could occur only if a sufficiently strong restraining layer outside of the cell prevented any swelling externally.

If the cavity remains constant in size, as shown in Figure 3.5c, the swelling of the gross wood for a given moisture change should be proportional to the volume of water absorbed and therefore to the density of the wood. A quantitative relationship should then be evident between the volumetric swelling of wood and its density.

If the cavity swells to some degree, the gross wood swelling is expected to be a maximum, as indicated in Figure 3.5d. If the lumen swelling is in the same proportion as that of the cell wall itself, then the gross wood swelling should be the same for all woods over a given moisture range. This is the situation which would be anticipated if all of the cell-wall layers had the same fibril orientation, that is, if the cell wall was homogeneous. Furthermore, the gross wood volumetric swelling and shrinkage would be given by equations (3.4) and (3.12) and would be equivalent therefore to those of the cell wall itself. This is the situation that exists when holes are drilled into a material such as wood. The over-all swelling and shrinking of the wood is the same as without holes, even though the "density" is now lower, based on over-all volume of the wood including the holes.

Of the three possibilities discussed above, the behavior of the individual woods appears to fit all three situations. In some woods the lumen appears to swell, in others it appears to shrink, and in others to remain essentially constant. As a general rule, however, the cell cavity appears to change only to a small extent during change of moisture content, unless collapse occurs during drying above the fiber-saturation point. This is a special case however, and will be discussed later in this chapter.

The general relationship that exists between total hygroscopic swelling and wood density can be derived if *it is assumed* that the cell cavity remains *constant* in size. In this case the total percent swelling S_f of the gross wood from dry condition to fiber saturation can be expressed as

$$S_f = 100 \, v'_{wf}/v_0 \tag{3.15}$$

in analogy with equation (3.3) where v'_{wf} is the volume of water sorbed by the fully swollen cell wall, and v_0 is the dry volume of the wood, including air space. Equation (3.15) can be converted into a form similar to equation (3.4) except that the dry specific gravity G_0 of the gross wood is now used rather than G'_0 for the dry cell wall. Thus equation 3.15 becomes

$$S_f = M_f G_0. \tag{3.16}$$

Similarly, it can be shown that the relationship of percent shrinkage s_f from M_f to dry condition can be written as

$$s_f = M_f G_g \tag{3.17}$$

where G_g is the specific gravity of the gross wood based on dry weight and green volume (or volume at fiber-saturation point M_f, which is presumed to be the same unless some collapse or stress-induced shrinkage occurs).

Equation (3.16) and (3.17) which are based on the assumption that the cell cavity remains constant in size when the moisture content of wood changes, predict that there should be a constant ratio between the total swelling S_f from dry to water-soaked condition, and the dry-volume specific gravity G_0, that this ratio should be equal to the fiber-saturation point M_f, and also that $M_f = s_f/G_g$. The hypotheses from which these equations were derived can be tested if sufficient data is available on woods covering a wide range of specific gravities. The data required are volumetric shrinkage s_f from the green or swollen condition to the dry state and also the moisture content M_f at the fiber-saturation point.

Stamm and Loughborough (1941) (see also Stamm 1964) reported that the mean ratio of s_f/G_g was 27 for 107 American hardwood species. This is approximately equal to the fiber-saturation point M_f, as is predicted by equation (3.17). There was considerable deviation from this mean value among the woods tested. For example, only 50 percent of them were within 11 percent of the mean value, and 75 percent were within 22 percent of the mean value.

The same authors found that the mean ratio of s_f/G_g was 26 for 52 different species of softwoods, 50 percent of which were within 10 percent of the mean value, and 75 percent within 19 percent of the mean value.

Greenhill (1936) (cited by Stamm and Loughborough 1941) found an average ratio s_f/G_g of 27 for 170 Australian species of wood. The scatter was larger, however, since only 50 percent of the species were within 21 percent of the mean, and 75 percent were within 33 percent of the mean.

Stamm and Loughborough reported s_f/G_g values of 25.4, 26.2, 24.0, and 24.8 with an average of 25.1 for loblolly pine boards whose corresponding values of G_g were 0.63, 0.54, 0.47, and 0.36. The slope of s_f against G_g for thin tangential sections of Douglas-fir containing different proportions of summerwood and springwood was 28.

Other data also have been used to support the concept that cell lumens in wood, on the average, remain constant during changes in wood moisture content. For example, Stamm (1935) has shown that lumen sizes of thin cross-sections of western white pine and of western hemlock remain practically constant and have the same permeability when air of different relative humidities is passed through them.

It is fortunate that the cell cavity does not change by the same percent as the cell wall because then the total value of s_f would be equal to $s_f' = M_f G_f'$ as given by equation (3.12). Thus, since G_f' is close to unity, from the calculations of Kellogg and Wangaard (1968) s_f would be essentially equal to M_f or near 25 to 30 percent, a much larger value than is normally found for the total volumetric shrinkage of wood.

The reason why cell lumens, on the *average*, tend to remain nearly constant in size during swelling or shrinking of the gross wood is probably related to the orientation of the fibrils in the three layers of the secondary wall (Figure 3.6). The central or $S2$ layer, the thickest layer, has its fibril orientation nearly parallel with the long axis of the cell and tends to swell transversely in proportion to the moisture change. The outer $S1$ and inner $S3$ layers, however, have their fibril orientation more nearly perpendicular to the cell axis and act to restrain the external and internal dimensional changes in the secondary wall. During swelling, for example, the outer $S1$ layer may act as a cross-band or restraining influence and may thus minimize the amount of external swelling. In order for this layer to swell appreciably the cellulose chains must be stretched to some extent, and since they are very strong along the length of the chain or fibril direction, they resist the stretching, thus reducing the external swelling. The inner $S3$ layer probably acts in much the same way to modify the swelling into the cell cavities. There are variations among different woods and even within the same wood in the relative thicknesses and fibril orientations of the various cell-wall layers. These variations may cause the deviations among individual woods from the general relationship given by equations (3.16) and (3.17).

Additional evidence indicates the restraining effect of the outer $S1$ layer in the observation of the excessive swelling that occurs in wood fibers when the outer layer is removed. This excessive swelling is called ballooning and may occur

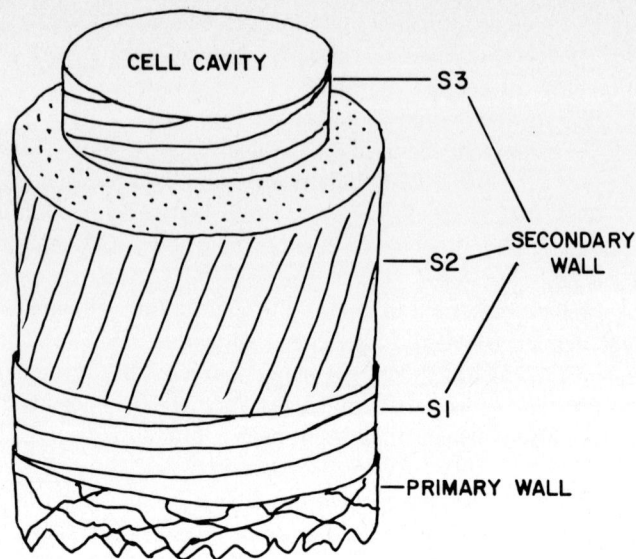

Fig. 3.6. Diagram showing the three layers of the secondary wall, the $S1$, $S2$, and $S3$ layers, and their fibril orientations.

after fibers have been exposed to excessive beating during the paper-making process.

More recent evidence to indicate the importance of the $S1$ and $S3$ layers in controlling swelling is given by Pollisco (1969), who observed excessive shrinkage when maple wood was treated with ammonia. Ammonia swells the cell wall more than water, and apparently the swelling stress is sufficient to crimp or destroy the integrity of the $S3$ layer, thus permitting swelling into the cell cavity. Parham (1971) has demonstrated the same effect in the $S1$ and $S3$ layers of loblolly pine.

The data given thus far to support the concept of cell-lumen constancy during moisture changes have not taken into account variations in the fiber-saturation point M_f among different woods. As equations (3.16) and (3.17) predict, the total shrinking or swelling between the green and dry condition should also be proportional to M_f.

Data by Higgins (1957) on fifteen different woods, including some woods native to the United States and also some foreign woods, give measured shrinkages s_f from green to dry condition, measured values of M_f, and also the specific gravities G_g based on dry weights and swollen volumes. These data are given in Table 3.3, together with the estimated values of s_f and M_f obtained by use of equation (3.17). The estimated values of M_f were obtained using the measured

Table 3.3. Specific Gravities, Volumetric Shrinkages, and Fiber-Saturation Points for Fifteen Woods Measured by Higgins (1957) and Calculated by Equation (3.17)

Species	G_0 (meas.)	G_g (meas.)	$M_f(\%)$ (meas.)	$s_f(\%)$ (meas.)	$M_f(\%) = s_f/G_g$ (calcul.)	$s_f(\%) = G_gM_f$ (calcul.)
W. Red Cedar	0.31	0.29	18.5	6.1	21.0	5.4
Redwood	0.37	0.34	22.2	7.8	23.0	7.5
Sitka Spruce	0.39	0.34	28.2	13.6	40.0	9.6
S.Y. Pine (#1)	0.44	0.39	29.0	12.8	32.8	11.3
S.Y. Pine (#2)	0.46	0.40	28.4	12.0	30.0	11.4
Teak	0.49	0.46	17.9	5.9	12.8	8.2
Mex. Mahogany	0.49	0.44	25.6	10.0	22.7	11.3
S.Y. Pine (#1-i)	0.56	0.50	16.1	6.9	13.8	8.0
Determa	0.56	0.50	23.3	10.4	20.8	11.6
S.Y. Pine (#2-i)	0.62	0.59	15.8	5.4	9.2	9.3
Coigue	0.63	0.50	31.0	18.8	37.6	15.5
Andaman Padouk	0.77	0.71	16.1	7.5	10.6	11.4
E. Ind. Rosewood	0.85	0.79	15.4	8.0	10.1	12.2
W. Ind. Locust	0.97	0.88	18.4	8.8	10.0	16.2
Ceylon Satinwood	1.09	0.94	19.1	13.6	14.5	17.9
Mean Value	0.600	0.538	21.67	9.84	20.59	11.12

values of s_f and G_g, using $M_f = s_f/G_g$. The estimated values of s_f were based on the measured values of G_g and M_f, using $s_f = G_gM_f$.

Figure 3.7 shows the regression of the estimated values against the measured values of M_f and also the measured against estimated values, together with the plotted points. Figure 3.7 also shows similar curves for the shrinkage s_f. It is clear that individual woods deviate considerably from the hypothetical relationship given by equation (3.17), although the mean regression curve is close to the hypothetical curve shown in the figures. This evidence supports the contention of Stamm that the cell cavity tends to remain constant, although there are considerable deviations in the behavior of individual woods.

A recent laminar sorption theory for the swelling of wood fibers (Stamm and Smith 1969) on the basis of a minimum swelling stress allows for a smaller *intra* laminar than *inter* laminar swelling. This minimizes the stress on the lumen wall thus allowing for constancy of the lumen or cell cavity size.

Maximum Possible Moisture Content of Wood

It has been shown previously that the maximum moisture content of the fully swollen cell wall is defined as the fiber-saturation point M_f or m_f. This is only a small portion of the maximum possible moisture content M_{max} of completely water-soaked wood because the cell cavities generally can hold more water than

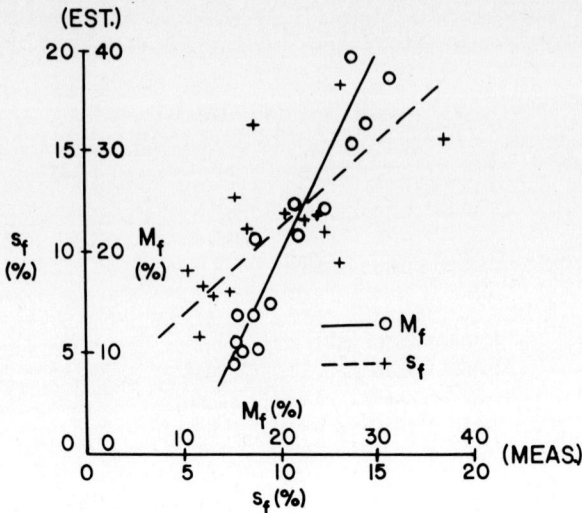

Fig. 3.7. Regression curves and plotted points of estimated against measured fiber-saturation points, M_f (%) and of estimated against measured volumetric shrinkages s_f (%) (from data of Higgins 1957).

the cell wall itself in most woods. We will now discuss the maximum possible moisture content which wood of a given specific gravity can attain when it is completely water soaked. The equations derived have been used to serve several purposes, one of which is a practical and rapid method of measuring the swollen-volume specific gravity G_f of wood, where G_f is taken to be equal to G_g, the specific gravity of green wood. Others have value in clarifying the swelling behavior of wood when it is exposed to water.

There are two related methods for calculating the maximum possible moisture content of wood. One of these methods is based on the assumption or theorem that the cell cavities remain constant in size during the sorption of moisture, and that M_f or m_f is known. Therefore it will be designated here as the theoretical constant-pore-volume method or simply as the theoretical method. The second method is entirely empirical since it makes no theoretical assumptions regarding pore-volume constancy. It will be designated here as the empirical method of maximum moisture content determination.

Theoretical Equation for Calculating Maximum Moisture Content

In this method the cell cavities are assumed to remain constant in size as the cell wall swells. The maximum quantity of water which can be held in the cell cavities as a function of the dry weight of the wood, designated as m''_{max}, is added to the maximum moisture content of the cell wall, designated as the fiber saturation point m'_f, equal to m_f if the air in the cell cavities is presumed to have negligible weight. The weight of water w''_w in the saturated cell cavities is equal to the product of the dry volume v''_0 of the cavities and the density ρ_w of water. The value of m''_{max} therefore is w''_w/w_0 where w_0 is the dry weight of the wood. Thus

$$m''_{max} = w''_w/w_0 = \rho_w v''_0/w_0. \tag{3.18}$$

However, v''_0 is given by

$$v''_0 = v_0(1 - G_0/G'_0) \tag{3.19}$$

where v_0 is the dry volume of the gross wood, and G_0 and G'_0 are the dry-volume specific gravities of the gross wood and of the cell wall. Therefore, from equations (3.18) and (3.19)

$$m''_{max} = (\rho_w v_0/w_0)(1 - G_0/G'_0) = (1/G_0)(1 - G_0/G'_0) \tag{3.20}$$

since the specific gravity G_0 of the dry wood is defined as $(w_0/v_0)/\rho_w$. Adding m''_{max} to m_f gives

$$m_{max} = m_f + (1/G_0)(1 - (G_0/G'_0)) \tag{3.21}$$

or

$$m_{max} = m_f + (G'_0 - G_0)/(G_0 G'_0) \tag{3.22}$$

or

$$M_{max} = M_f + 100(G'_0 - G_0)/(G_0 G'_0). \tag{3.23}$$

Equations (3.21), (3.22), or (3.23) can be used to calculate the maximum possible moisture content if M_f, G'_0, and G_0 are known and if the cell cavity remains constant when the wood increases from the dry condition to saturation.

Empirical Equation for Calculating Maximum Moisture Content

The second method for finding M_{max} for wood is based on an empirical relationship and requires no theoretical assumptions with regard to the behavior of the cell-cavity volume during water sorption. It is derived by recalling that the total volume v_w of water in completely saturated wood is equal to the difference between the fully swollen volume v_f of the gross wood and the dry volume v'_0 of the cell wall, and that the weight w_w of water is equal to $\rho_w v_w$. Thus,

$$w_w = \rho_w v_w = \rho_w(v_f - v'_0). \tag{3.24}$$

Therefore

$$m_{max} = w_w/w_0 = (\rho_w v_f/w_0) - (\rho_w v_0'/w_0) \qquad (3.25)$$

$$m_{max} = (1/G_f) - (1/G_0') = (G_0' - G_f)/(G_f G_0') \qquad (3.26)$$

$$M_{max} = 100\,[(1/G_f) - (1/G_0')] = 100(G_0' - G_f)/(G_f G_0') \qquad (3.27)$$

which is the equation used by Smith (1954).

Comparison of Results Obtained by the Two Equations

It should be noted that in both equations the dry-volume specific gravity of the cell wall G_0' is based on the value obtained by water displacement.

It is of interest to compare the results obtained for M_{max} by the two equations. If the results are identical then the hypothesis of constant cell-cavity volume is indicated, provided the fiber-saturation point M_f is correct. The data obtained by Higgins (1957) shown in Table 3.3 gives the opportunity of checking this relationship directly. The only missing factor is G_0', the specific gravity of the dry cell wall. This does vary somewhat among different woods, as Table 3.1 indicates. The mean value of V_{0w}' (the reciprocal of G_{0w}') based on water displacement as found by Kellogg and Wangaard (1969) for eighteen United States woods, was 0.650 cc/g ($G_{0w}' = 1.538 \approx 1.54$). Using this value and the individual values of G_0, G_f, and M_f given in Table 3.3 from Higgins' data, the values of M_{max} can be calculated for any wood using both equations (3.23) and (3.27).

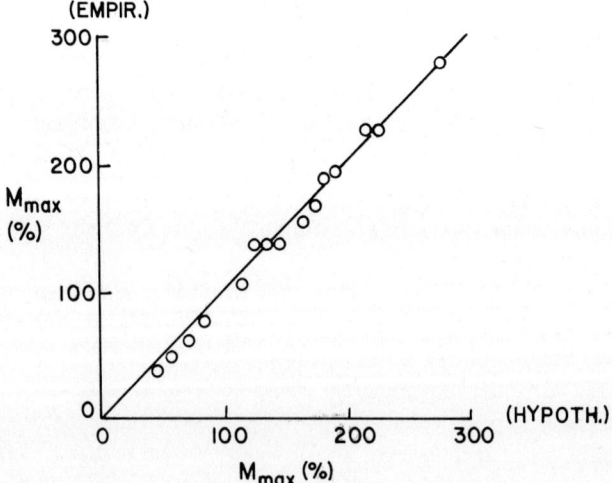

Fig. 3.8. Comparison of maximum moisture contents M_{max} (%) calculated from the empirical and hypothetical equations based on data from Higgins (1957).

For example, substitution of the values given in Table 3.3 for western red cedar give $M_{max} = 18.5 + 100(1.54 - 0.31)/[(0.31)(1.54)] = 276$ percent from equation (3.23) and $M_{max} = 100(1.54 - 0.29)/[(0.29)(1.54)] = 280$ percent from equation (3.27). This procedure was carried out for each of the fifteen woods listed in Table 3.3 and the results plotted in Figure 3.8. If the assumption of constant cell-cavity volume with moisture change is valid and if the values of M_f are correct the two equations should give identical values for M_{max} and the points should fall in a straight line which passes through the origin and has a slope of unity.

Equation 3.27, the empirical equation for finding M_{max}, can also be used to calculate G_f if G'_{0w} and M_{max} are known, as Keylwerth (1954) has shown. This is a useful procedure for small samples such as those obtained from increment borings. Four possible sources of error are pointed out by Smith (1954), who has also used this technique successfully. These are: "(1) obtaining the absolute maximum saturation with water; (2) obtaining the soaked weight of the sample in air; (3) assuming that the density of wood substance for a species is constant; and (4) the possibility of extraneous matter in the sample."

Another possible method of using equation (3.27) is to calculate the specific gravity G'_{0w} of dry-wood substance based on water displacement. In this case M_{max} and G_f are measured experimentally. If the sample is one cubic centimeter or larger G_f can be measured readily by water displacement and M_{max} can be measured if all of the air spaces in the specimen can be displaced by water. This is difficult to do for impermeable woods, and in any event it is usually necessary alternately to apply vacuum and pressure (under water) to a sample until no more air bubbles appear under vacuum.

Petty and Preston (1969) have carried out experiments on sapwood and heartwood of seven woods in the form of cylinders 2 cm in diameter and 11 cm long parallel to the grain to determine what fraction δ of the air in the void volume could not be removed from the wood by evacuation. They found that the impervious fraction δ for sapwood of most of the seven woods ranged from 0 to 0.05 except for *Eucalyptus maculata* which was from 0.33 to 0.43. The heartwood values of δ ranged from 0.02 to 0.55 except for *Eucalyptus maculata* in which case it was 0.74 to 0.95. They pointed out the errors which might be involved in the maximum moisture-content method for measuring G'_0 for the cell wall of some of the impervious woods. They used glass beads of 1 mm diameter as a displacement medium to measure the gross volume of the wood samples, rather than using a liquid medium.

Initial Shrinkage and Dimensional Stability

In the previous paragraphs we have discussed only the maximum volumetric shrinking and swelling of wood between the dry and fully swollen condition.

Normally wood in use is not subjected to such extreme moisture changes, and the question arises as to the degree of dimensional changes which take place in wood volume—first when it dries initially from the green condition to some desired moisture content and second during the normal cyclic changes in environmental conditions to which it is exposed. The latter changes have been termed "movement" by Stevens (1963) to distinguish them from the shrinkage that takes place during the initial drying. Before discussing these two factors in detail we will first consider the relationships among volumetric dimensional changes and those in the longitudinal, radial, and tangential directions. For convenience we will discuss these relationships in terms of wood shrinkage, although they also apply to wood swelling and to cyclic changes or movement.

Volumetric shrinkage measurements alone are not sufficient to describe the shrinkage which takes place in wood during drying because wood is orthotropic with respect to shrinkage as well as in most of its other physical properties. The reasons for the differential shrinkages in the three primary directions, longitudinal, radial, and tangential, are discussed below. At this point we will only state that longitudinal shrinkage s_l is least, usually in the order of 0.1 to 0.3 percent from fiber saturation to the completely dry condition in normal wood. Total radial shrinkage s_r ranges from about 3 to 6 percent and tangential shrinkage s_t from about 6 to 12 percent for most woods over the same moisture range.

Volumetric shrinkage s_v is slightly less than the sums of the three directional components. The exact expression (Greenhill 1936) for a prismatic sample is

$$s_v = 100\,[1 - (1 - 0.01\,s_l)\,(1 - 0.01\,s_r)\,(1 - 0.01\,s_t)]$$
$$= s_l + s_r + s_t - (0.01)\,(s_l s_r + s_r s_t + s_t s_l) + (0.0001)\,(s_l s_r s_t) \qquad (3.28)$$

where the shrinkages are given in percent. A slightly less exact but practically identical equation can be obtained from equation (3.28) by eliminating the terms in which the longitudinal shrinkage s_l occurs as a multiple with s_r or s_t or both. Since s_l is so small for wood these terms become negligibly small even when s_r and s_t are large. This simplification reduces equation (3.28) to

$$s_v \approx s_l + s_r + s_t - 0.01\,s_r s_t \qquad (3.29)$$

which is sufficiently accurate within the range of s_r and s_t values normally found for wood. If the last term involving the product $s_r s_t$ is neglected, the equation reduces further to the simple form

$$s_v \approx s_l + s_r + s_t \qquad (3.30)$$

or if s_l is completely negligible compared with s_r and s_t

$$s_v \approx s_r + s_t. \qquad (3.31)$$

The magnitude of the errors involved in using equation (3.30) rather than the exact equation (3.28) or the nearly exact equation (3.29) depends on the magnitudes of s_r and s_t. Ordinarily s_t is approximately twice s_r, the ratio of s_t/s_r commonly being called the T/R ratio. Table 3.4 shows the values of s_v calculated

Table 3.4. Comparison of Percent Volumetric Shrinkage s_v Obtained Using the Approximate Equation (3.29) and the Exact Equation (3.28) for T/R Ratios of 1, 2, and 3

Approximate Equation (3.29)	Exact Equation (3.28)		
Calculated Value of s_v (%)	$T/R = 1.0$	$T/R = 2.0$	$T/R = 3.0$
0.00	0.00	0.00	0.00
3.00	2.98	2.98	2.98
6.00	5.91	5.92	5.93
9.00	8.80	8.82	8.85
12.00	11.64	11.68	11.73
15.00	14.44	14.50	14.58

from equations (3.28) and (3.29) over the range of s_v normally found for wood, assuming T/R ratios of 1, 2, and 3, and that the longitudinal shrinkage s_l is small compared with s_r and s_t. It is clear that the difference between the calculated volumetric shrinkages based on the exact and approximate equations increases as the shrinkage increases and as the T/R ratio decreases toward unity.

Having discussed the relationships between volumetric and directional shrinkages we return now to a more general treatment of dimensional changes associated with moisture changes after initial drying. As is the case with shrinkage it is necessary to distinguish between volumetric and directional dimensional changes. The symbol X_v will be used to represent the coefficient of volumetric moisture expansion or of volumetric swelling defined as follows:

$$X_v = (1/v)(dv/dm) \tag{3.32}$$

where v is the volume and dv/dm is the change in volume per unit change in fractional moisture content. Similar expressions for directional moisture expansion or swelling coefficients are

$$X_l = (1/l)(dl/dm) \quad \text{(longitudinal)} \tag{3.33}$$

$$X_r = (1/r)(dr/dm) \quad \text{(radial)} \tag{3.34}$$

$$X_t = (1/t)(dt/dm) \quad \text{(tangential)}. \tag{3.35}$$

Figure 3.9, which shows a hypothetical idealized curve of wood volume v as a linear function of moisture content m, can be used to show how X_v varies with moisture content m. The slope dv/dm is constant but the reference volume v de-

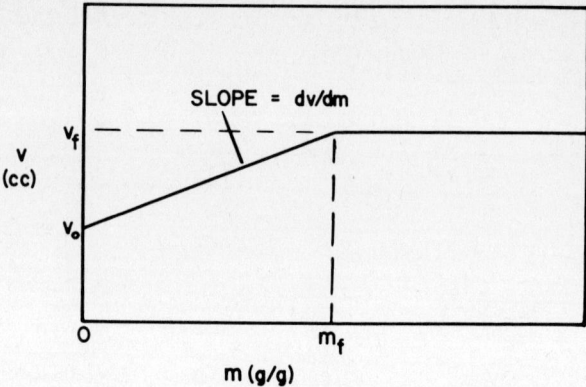

Fig. 3.9. Idealized linear curve of wood volume v (cc) against wood moisture content m (g/g).

pends on m. Therefore X_v is a maximum at $m = 0$ and a minimum at $m = m_f$. It can be shown that X_{v0}, the value of X_v at $m = 0$, is related to S_f, the total percent volumetric swelling from the dry condition to M_f, as follows

$$X_{v0} = S_f/M_f \tag{3.36}$$

since $X_{v0} = (1/v_0)(dv/dm) = (\Delta v/v_0)/\Delta m = S_f/100\,M_f = S_f/M_f$. Similarly it can be shown that

$$X_{vf} = s_f/M_f \tag{3.37}$$

where X_{vf} is the value of X_v at M_f, and s_f is the total percent volumetric shrinkage from M_f to the dry condition.

For the special case where the cell-cavity volume remains constant with moisture change, equations (3.36) and (3.37) can be simplified by use of equations (3.16) and (3.17), and become

$$X_{v0} = G_0 \tag{3.38}$$

and

$$X_{vf} = G_f = G_g. \tag{3.39}$$

It is also possible to define average swelling coefficient $\overline{X}$ over the hygroscopic moisture range by taking the means of X_{v0} and X_{vf}. Thus

$$\overline{X}_v = (X_{v0} + X_{vf})/2 \tag{3.40}$$

which for the special case of constant cell-cavity volume given by equations (3.38) and (3.39) becomes

$$\overline{X}_v = (G_0 + G_f)/2. \tag{3.41}$$

Similar relationships can be obtained for the directional swelling coefficients X_l, X_r, and X_t except for those involving G_0 and G_f which apply specifically to volumetric changes. These coefficients are analogous to thermal-expansion coefficients and might be called "hygro-expansion" coefficients. They can be expressed as fractional changes in volume or length per unit fractional change in moisture content m, or in terms of percent volume or length change per percent change in moisture content M, or in the hybrid system of fractional dimensional change per percent moisture content change.

It has been assumed throughout the preceding discussion that the dimensional changes in wood are linear with moisture changes in the hygroscopic range. This is an approximation which is made for convenience and is rarely observed when careful measurements are made. Even when there are no cell-cavity changes or drying stresses caused by moisture gradients, one would expect to have slightly nonlinear dimension-moisture relationships for at least two reasons. First, if there are voids in the cell wall which are penetrated by water, as Weatherwax and Tarkow (1968) propose, one would expect that the water which penetrates these voids would not contribute to the swelling of the cell wall (Figure 3.10).

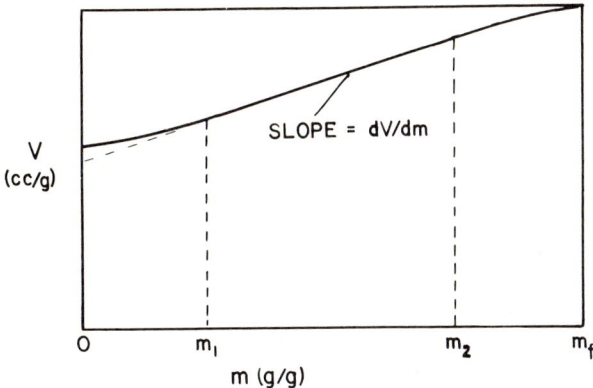

Fig. 3.10. Curve of specific volume V (cc/g) of wood against moisture content m (g/g) showing regions of nonlinearity.

Second, if the apparent specific volume of the sorbed water increases with increasing wood moisture content, as Stamm (1964) has proposed, then the volumetric swelling coefficient based on dry dimensions should increase with moisture content because the same quantity of water taken up at a low moisture content has a lower volume than that taken up at a higher moisture content.

Keylwerth (1964) gives curves for a number of woods, of radial r, tangential t, and volumetric v swelling as functions of wood moisture content during adsorption. All of these curves are slightly sigmoid as is shown in Figure 3.11 for European birch. During earlier stages of adsorption the rate of swelling dv/dM is

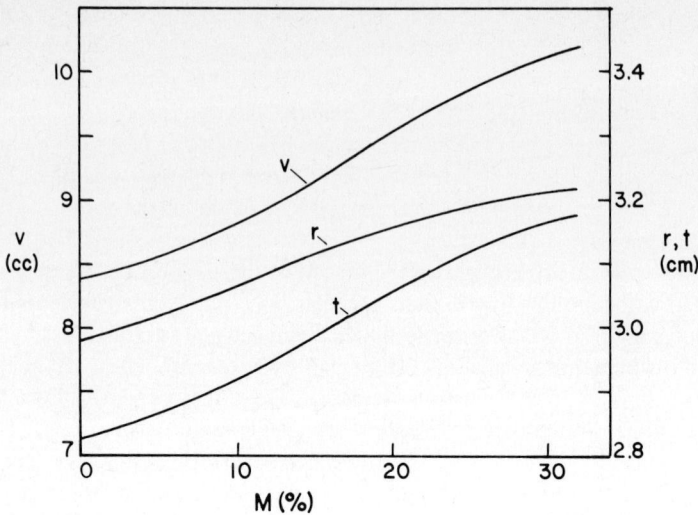

Fig. 3.11. Experimental curves showing volumetric v (cc), radial r (cm), and tangential t (cm) dimensional changes with wood moisture content for European birch (adapted from Keylwerth 1964).

relatively small and increases with increasing wood moisture content up to about 4 or 5 percent. This may be caused by the increase in apparent specific volume V_w' (decreasing density ρ_w') of the sorbed water with increasing moisture content because of the relatively high packing density of the sorbed water at low wood moisture contents, as discussed previously. It is also conceivable that some of the low-volume swelling at these low moisture contents is caused by water molecules being sorbed on the internal surfaces of the microvoids which have been postulated to exist in the cell wall (Weatherwax and Tarkow 1968), and therefore little volume change takes place in the cell walls. Above about 5 percent moisture content the increase is essentially linear until about 25 percent, above which swelling rate decreases. This may be caused by the filling of existing microvoids at such high moisture contents by capillary condensation as predicted by the Kelvin equation. It may also be caused by swelling stresses analogous to the shrinkage stresses discussed below. Again, less volume change per unit mass of water is involved.

Keylwerth (1964) has replotted the curves shown in Figure 3.11 in terms of their slopes, divided by the dry volume v_0 or dry dimensions r_0 and t_0 with the results shown in Figure 3.12. These are really curves of $X_{v0} = (1/v_0)(dv/dm)$ for volumetric swelling, $X_{r0} = (1/r_0)(dr/dm)$ for radial swelling, and $X_{t0} = (1/t_0)(dr/dm)$ for tangential swelling, as defined by equations (3.32), (3.34), and (3.35), based on dry dimensions. It is clear from the curves of Figures 3.11 and 3.12 that swelling is essentially linear over most of the moisture range except

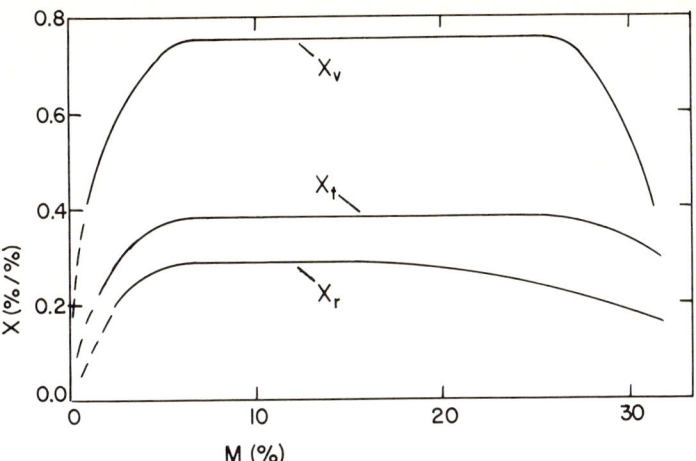

Fig. 3.12. Curves of volumetric X_v, radial X_r, and tangential X_t hygroexpansion coefficients based on dry dimensions for European birch (adapted from Keylwerth 1964).

near zero and near m_f. If the ratio of the observed value of X_{v0}, divided by the dry-volume specific gravity G_0, is plotted against m for a given wood, one would anticipate a value of unity if the cell cavity remains constant during swelling and if swelling is proportional to the volume of water taken up. Keylwerth (1964) has plotted such curves for a number of woods, with the results shown in Figure 3.13. It is clear from the curves that some woods show swelling of the cell cavity (those which show a ratio greater than unity), and others show some shrinkage of the cell cavity (those which show a ratio less than unity). It is also clear that most of the woods show nonlinear swelling at the lower and higher portions of the moisture range, as was the case with birch shown in Figure 3.12. Keylwerth also showed that there are considerable variations within a single species from different sources with respect to swelling behavior.

There is a more serious nonlinearity in the initial shrinkage which occurs with moisture loss when wood dries from the green condition. This is illustrated in Figure 3.14, adapted from Kauman (1964), which shows the shrinkage-moisture content relationships for a collapse-prone wood dried under three conditions. The curve AB represents the normal shrinkage based on green dimensions which would result in wood which dries from the green condition without collapse. The curvilinearity between 20 and 40 percent moisture content may represent the effect of stresses caused by moisture gradients. For example, the outer surfaces of the wood dry first and tend to shrink, forcing the more moist but weaker wood in the interior of the piece to shrink, even though the mean moisture content of the entire sample is somewhat above fiber saturation.

The curve CDE shows the shrinkage for a sample which collapses severely as it

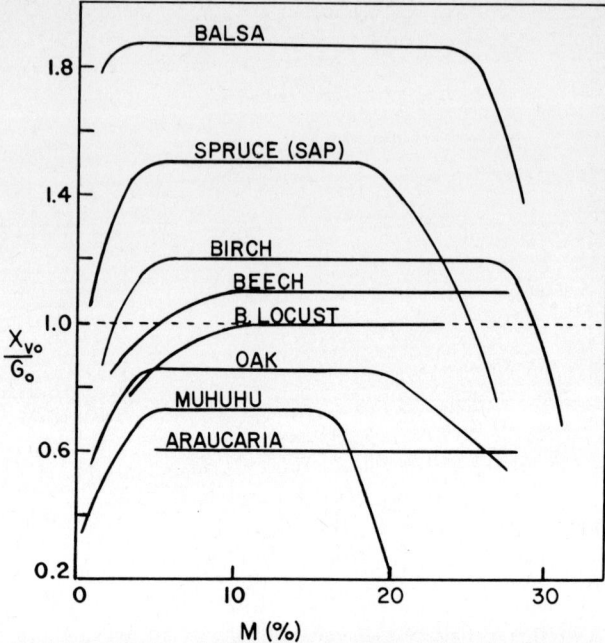

Fig. 3.13. Curves of volumetric hygroexpansion coefficients X_{v0} divided by the specific gravity G_0 based on dry volume for several woods (adapted from Keylwerth 1964).

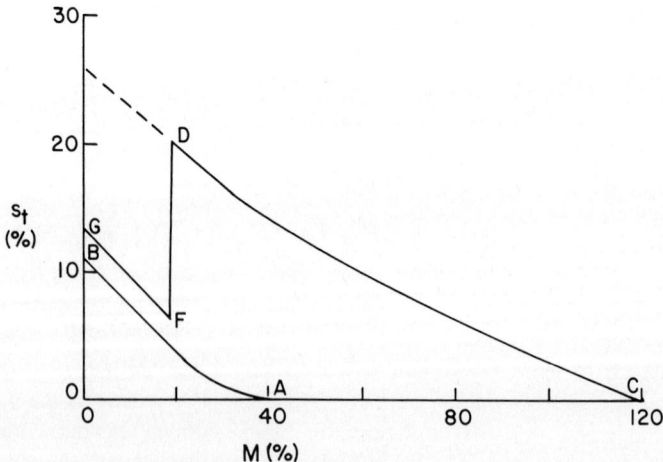

Fig. 3.14. Relationship of tangential shrinkage s_t (%) to wood moisture content M (%) for a collapse-prone wood as a function of treatment history (adapted from Kaumann 1964).

dries. Note that it begins to shrink long before it reaches the fiber-saturation point of approximately 30 percent. A conditioning treatment originally ascribed to Tiemann (1944) which restores the wood almost to its collapse-free condition may be applied when the moisture content reaches 15 to 20 percent, represented by point D. This treatment consists of raising the temperature and humidity of the dry kiln for a brief time. This results in raising the water vapor pressure of the wood and also its temperature, as well as adding a slight amount of water to the wood. The increased vapor pressure in the cell cavities, coupled with the softening of the wood by heat and moisture, effectively restores the collapsed cells almost to their original dimensions, shown by point F. Subsequent drying follows the curve FG which shows somewhat higher shrinkage than the collapse-free curve AB.

Kübler (1970) has shown that heat recovery of wood with excessive initial shrinkage or collapse after drying of red oak samples heated in sealed vaportight compartments can be accomplished at moisture contents as low as 11 percent. Heating sealed samples in the moisture range between 11 and 15 percent gave the most favorable recovery of this excessive shrinkage.

A detailed discussion of the mechanism of collapse in wood during drying is not offered in this book since it is covered adequately by Siau (1971) in the first book in this series. It should be pointed out, however, that it is believed to be caused primarily by the liquid capillary tension in the cell-cavity water as liquid water is evaporated from the cells near the drying surface of the wood during the early stages of drying. The magnitude of the tension is determined by the Kelvin equation or by the osmotic-pressure or swelling-pressure equation (Chapter 1), depending on whether we are relating capillary radius or relative vapor pressure to the liquid tension forces.

Anisotropy in Shrinking and Swelling

It has already been pointed out that wood is anisotropic with respect to its dimensional changes with changing moisture content. Normal wood is most stable dimensionally in the parallel-to-grain direction and least so in the tangential direction. Fortunately, wood also possesses its highest mechanical strength when external forces are applied so that the stresses in the wood are parallel to the grain. For example, wooden beams in the floors of buildings are loaded in such a way that the internal compression and tensile stresses are maximized along the grain. Likewise the hygroscopic dimensional changes are smallest in the long direction of the beams, thus reducing the length changes which might otherwise take place. The same factor of small longitudinal hygroscopic movement is also important when wood is used for sheathing or flooring since the larger dimension is always parallel to the grain. We will therefore first consider the longitudinal shrinkage of wood and the factors which affect it before discussing transverse shrinkage in the radial and tangential directions.

Longitudinal Shrinking and Swelling

Longitudinal or axial shrinkage in wood is generally one or two orders of magnitude less than transverse shrinkage. It usually ranges from 0.1 to 0.3 percent when normal mature wood dries from the green to ovendry condition. Sometimes negative shrinkage occurs along the grain; that is, the wood may be slightly longer in the air-dry condition than in the green condition (Hann 1969). Therefore, normal mature wood from the outer portions of straight older trees shrinks so little along the grain that there is no shrinkage problem in use. However, the wood known as juvenile wood, which comes from young trees or from near the pith of older trees, may have appreciably higher longitudinal shrinkage than those given above. This is believed to be the result of the difference between the fibril angles in the $S2$ layer in the cell walls of juvenile wood and those of mature wood. The increased use in recent years of young trees which contain a substantial amount of juvenile wood has increased interest in the longitudinal shrinkage of wood. Also, the increased use of reaction wood, designated as compression wood in softwoods and tension wood in hardwoods, both of which also show excessive longitudinal shrinkage (Panshin and de Zeeuw 1970), focusses further attention on this factor.

Koehler (1946) lists three conditions of use under which longitudinal shrinkage may be considered excessive. These are:

1. When it causes undue shortening in length so as to throw adjoining members into serious disalignment, as in posts and columns, . . . and opening of butt joints, as in flooring and siding.

2. When it causes crook or cross breaks in lumber or waviness in veneer due to unequal shrinkage.

3. When it causes stresses in lumber that result either in pinching of the saw in ripping or crosscutting, or in longitudinal cracking of boards in ripping or machine planing.

Several theories have been proposed to explain the factors which affect longitudinal shrinkage of wood. These are usually based on the fibril angles in the $S2$ layers of the cell walls of the longitudinal cells. Cockrell (1946) proposed a simplified lattice structure to explain longitudinal shrinkage, with some success. A more quantitative theory is that of Barber (1968), which is a refinement of the earlier theory of Barber and Meylan (1964). These theories are based on models which consider the cell wall of wood longitudinal cells to consist of an amorphous hygroscopic matrix in which are embedded parallel crystalline microfibrils which act to restrain swelling or shrinking in the direction parallel to their axes.

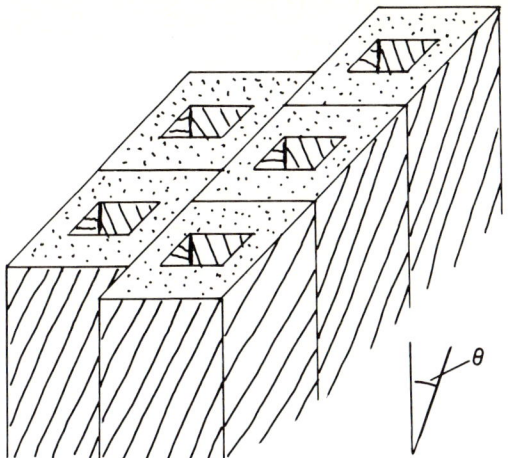

Fig. 3.15. Simplified cell-wall model showing fibril angle θ (adapted from Barber and Meylan 1964).

In the earlier (1964) model all of the fibrils are assumed to be oriented at the same angle θ to the long axis of the cell and the cells are assumed to be square in cross-section (Figure 3.15). Furthermore, since the adjoining walls of adjacent cells have fibrils of the same orientation θ but of opposite sense, the amorphous hygroscopic material being continuous throughout both cells, there is no tendency for the individual cells to twist as they swell. In the later (1968) model the cells are assumed to be circular in cross-section, and a thin elastic constraining sheath, unaffected by water, is presumed to surround each thickwalled cylindrical cell. It is convenient here to discuss first the results obtained with the simplified (1964) model after which the modified results obtained by the later (1968) model will be considered.

The principles involved in anisotropic swelling can best be visualized by reference to Figure 3.16, adapted from Barber and Meylan (1964). In each of the diagrams, **a**, **b**, and **c**, an inner square represents the original dry dimensions of a section of the cell wall as seen from the cell cavity, with the long axis of the cell vertical. The fine vertical solid line in **a** and the diagonal fine solid lines in **b** and **c** represent microfibrils which resist length changes and therefore partially restrain the amorphous region from swelling in the direction parallel to their axes. The large outer square outlined with the broken line represents the swollen shape and size of the cell-wall section if there were no microfibrils to restrain swelling. They are of the same dimensions in **a**, **b**, and **c** and indicate isotropic swelling.

Referring to **a**, which shows the microfibrils parallel to the cell axis ($\theta = 0°$), there is less swelling along the cell axis than in the transverse direction because

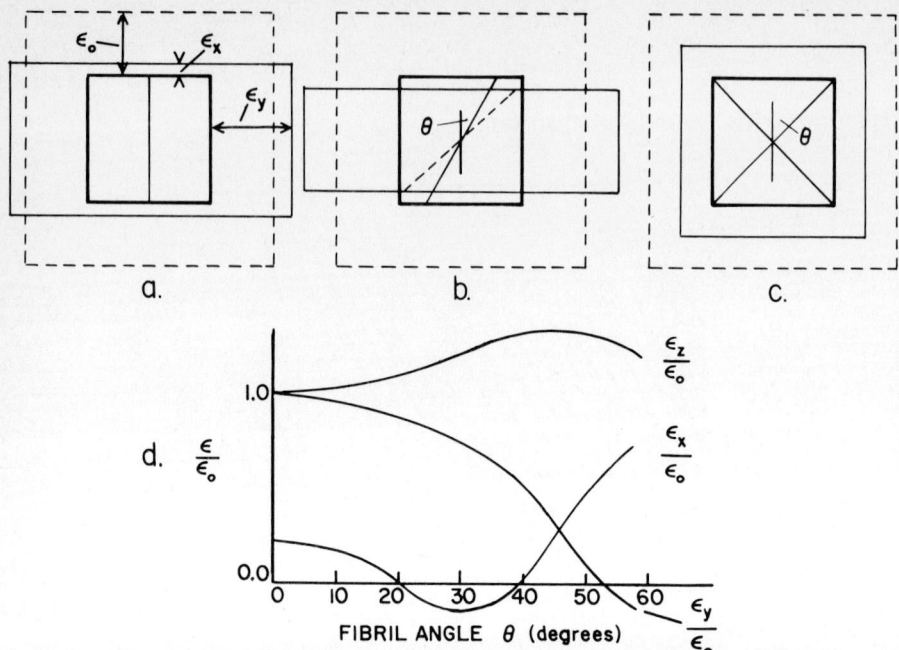

Fig. 3.16. Diagrams *a*, *b*, and *c* show anisotropic swelling for fibril angles θ of $0°$, $30°$, and $45°$. Diagram *d* shows the strain ratios ϵ/ϵ_0 as functions of fibril angle θ (adapted from Barber and Meylan 1964).

of the restraining effect of the microfibrils in the longitudinal direction. The over-all swollen shape of the cell section is now rectangular with only a small component of swelling along the cell axis because of the high tensile strength of the microfibrils. The transverse swelling, however, is larger than for the isotropic case because the Poisson effect tends to keep the total swollen volume nearly equal to that for the isotropic case. One would expect some reduction in total swelling, however, because of the Barkas effect described in Chapter 4.

In diagram **b** the microfibril angle θ for the dry wood is $30°$, the angle between the solid diagonal line and the cell axis. After swelling, the microfibril has shifted in position to that shown by the broken line. The microfibril may be slightly stretched because of the tensile stress induced by the swelling of the matrix material. However, because of the increase in fibril angle during swelling the vertical component of the microfibril length actually decreases while the horizontal or transverse component increases. The swollen shape of the section is as shown by the rectangular outline in **b**. It has actually decreased in the axial direction of the cell, resulting in a negative swelling along the cell axis. There is also a larger swelling in the transverse direction than in the case shown in **a** where the angle θ is $0°$.

When the fibril angle θ is 45°, as is shown in diagram c, the swelling tends to be isotropic in the longitudinal and transverse directions. The swelling is lower than for the case shown by the broken line outline where no microfibrils are present but the shape is the same.

It should be pointed out that there tends to be excessive swelling in the direction of the cell-wall thickness (perpendicular to the plane of the diagrams) in all three cases, a, b, and c, because there is no restraint in this direction. However, when the cells are square the increase in thickness in one wall is equalled by the increase in thickness of the adjoining wall at right angles to it. Therefore, the over-all cross-sectional shape of the cell remains square after swelling for all fibril angles, based on this model.

Figure 3.16d, also adapted from Barber and Meylan (1964), shows three curves which illustrate qualitatively how the ratio ϵ/ϵ_0 of the expected swelling ϵ in each of the three principle directions when microfibrils are present to the isotropic swelling ϵ_0 which would occur in the matrix if the restraining effect of the microfibrils was absent. Looking first at ϵ_x/ϵ_0 where ϵ_x is the swelling in the vertical direction, parallel to the cell axis, it appears that ϵ_x decreases with increasing fibril angle θ until 30°, above which it increases. Between $\theta = 20°$ to 40° the swelling is somewhat negative, indicating that longitudinal shrinkage occurs as the wood gains moisture.

The transverse swelling ratio ϵ_y/ϵ_0 begins at a maximum value when the fibril angle is zero and decreases slowly up to about 30° at which point it decreases rapidly, becoming equal to ϵ_x/ϵ_0 at $\theta = 45°$, as is indicated in Figure 3.16c. The thickness swelling ratio ϵ_z/ϵ_0 is equal to the transverse swelling ratio ϵ_y/ϵ_0 at $\theta = 0°$, but increases with increase in θ while ϵ_y/ϵ_0 decreases.

The improved cell model of Barber (1968) shown in Figure 3.17 differs from the earlier model in two respects; it is circular in cross-section and also provides for a thin constraining sheath outside the cylinder which acts to reduce transverse swelling. The cell wall therefore is considered to consist of two layers—the thin constraining sheath representing the $S1$ layer and the thick dominant layer of thickness $r_2 - r_1$, representing the $S2$ layer (Figure 3.6), where r_1 and r_2 are the inner and outer radii of the cell wall, essentially equivalent to the same radii for the $S2$ layer. For convenience in deriving his general equation Barber also makes the simplifying assumption in his cylindrical model that there are two sets of microfibrils in the $S2$ layer winding in opposite directions but at the same angle θ from the cell axis. This overcomes the twisting tendency which exists in a single isolated cell and has essentially the same effect as having two adjacent wall with microfibrils oriented in opposite senses (Figure 3.15) and connected together by the middle lamella as they are in wood.

The equations used by Barber (1968) and by Barber and Meylan (1964) are derived in their papers and will not be given here. However, a graphical presentation of some of Barber's calculated results is shown in Figure 3.18, adapted from Barber, in the form of six curves. All of the curves are calculated on the assump-

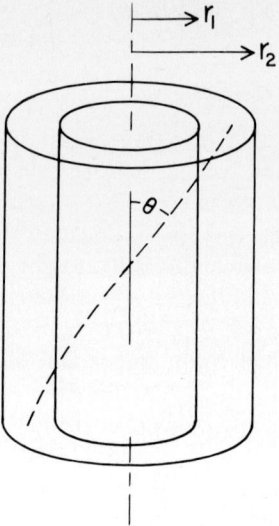

Fig. 3.17. Cylindrical cell model of Barber (1968) showing inner and outer radii r_1 and r_2 and fibril angle θ.

tion that the rigidity of the restraining sheath is one percent of that of the $S2$ layer in the direction parallel to the microfibrils. Essentially this means that the sheath contains microfibrils oriented perpendicularly to the cell axis and that these have a total stiffness in restraining extension of only one percent compared with the total stiffness of those in the $S2$ layer. Presumably this small percent is based on the small volume of microfibrils in the thin sheath.

Two of the curves in Figure 3.18 are qualitatively similar to the ϵ_x/ϵ_0 curve for longitudinal swelling shown in Figure 3.16d, and are labeled in the same way. The two curves in Figure 3.18 differ from each other in the ratio E/S, where E is the effective stiffness of the microfibrils in resisting elongation by the swelling matrix and S is the rigidity or shear modulus of the swelling matrix proportional to its resistance to shear deformation. It is clear from the figure that the relative longitudinal swelling at low fibril angles is smaller when the relative stiffness ratio E/S is larger. This is as expected because the stiffer the microfibrils are compared with the swelling matrix, the greater effect they will have in reducing longitudinal swelling at low angles of θ. It is also clear that the calculated longitudinal swelling ϵ_x is only a small fraction of the swelling ϵ_0 expected for the isotropic matrix at small values of θ. As the angle θ increases from zero the longitudinal swelling approaches zero and actually becomes negative, reaching a minimum near $30°$ and then increasing sharply above this point.

A second pair of curves in Figure 3.18 shows the ratio ϵ_2/ϵ_0, where ϵ_2 is the calculated external transverse swelling (increase in radius r_2) when microfibril

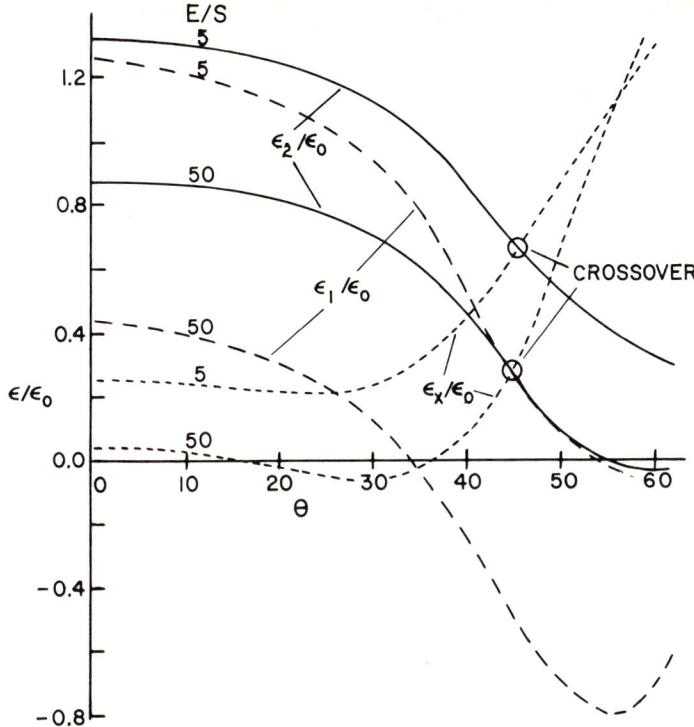

Fig. 3.18. Strain ratio ϵ/ϵ_0 curves of Barber for longitudinal ϵ_x/ϵ_0, transverse internal cell dimensions ϵ_1/ϵ_0, and external cell dimensions ϵ_2/ϵ_0 as functions of fibril angle θ for two ratios E/S of the relative stiffness of restraining sheath to the swelling cell-wall material (adapted from Barber 1968).

restraint is considered and ϵ_0 is the external transverse swelling expected for the isotropic matrix. Each of these curves is also for a different E/S ratio. It is clear from the figure that for low fibril angles the calculated value of swelling ϵ_2 is appreciably larger than the isotropic swelling ϵ_0. It remains essentially constant up to a fibril angle near $25°$ after which it decreases rapidly, crossing over the rapidly rising curves of ϵ_x/ϵ_0 at a fibril angle θ near $45°$.

The third pair of curves shows the ratio ϵ_1/ϵ_0, where ϵ_1 is the swelling of the cell cavity or the increase in radius r_1 in Figure 3.17. In most cases the relative swelling ϵ_1 is smaller than that of ϵ_2, the external cell swelling. It may approach zero for high ratios of E/S when the restraining sheath is sufficiently rigid. This would be the case found for many woods in which the cell cavity appears to remain nearly constant in size during moisture changes, as discussed above.

Experimental confirmation of the effect of fibril angle on the longitudinal and transverse shrinkage, including "crossover" of the two shrinkage components at

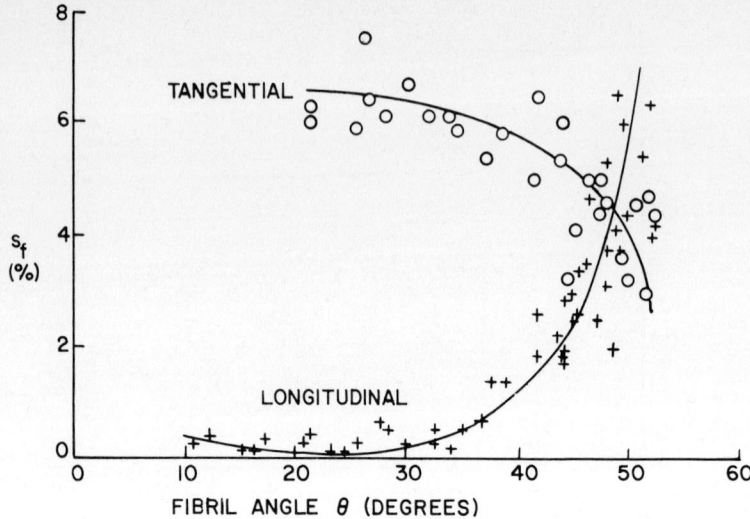

Fig. 3.19. Experimental curves of the longitudinal and tangential shrinkages of *Pinus jeffreyi* as function of fibril angle θ (adapted from Meylan 1968).

large fibril angles, is shown in Figure 3.19, taken from Meylan (1968). Here are shown the relationship between longitudinal and tangential shrinkage and micro-fibril angle for *Pinus jeffreyi*. The crossover point occurs at a microfibril angle between 45° to 50° at which the two shrinkage components are equal. The shapes of the curves are similar to the theoretical curves derived by Barber and shown in Figure 3.18. The exact shape depends on factors which are difficult to evaluate quantitatively but the general agreement is good.

Barrett, Schniewind, and Taylor (1972) extended the model of Barber (1968) to account more fully for the sorption behavior of the constituent components of the cell wall, including cellulose, hemicellulose, and lignin. Furthermore, they used relative humidity rather than moisture content as the primary variable to account for differences in the sorption isotherms of these constituents.

Radial and Tangential Shrinkage

It has already been pointed out that the longitudinal shrinkage of wood is considerably smaller than transverse shrinkage except under those rare circumstances when the fibril angle θ is excessively steep. It also has been stated that there is considerable anisotropy in transverse swelling itself in that radial shrinkage is usually about half the tangential shrinkage over a given moisture range. This behavior has probably been observed in wood so long as it has been used by man and has been responsible for much of the warping of shape associated with wood drying and also with its subsequent use. We have already referred to the T/R ratio as being the ratio of tangential shrinkage s_t to radial shrinkage s_r, and have indicated that it varies among woods, averaging close to 2 for many woods. Before discussing the mechanisms which have been proposed to explain transverse anisotropy we will consider some of its practical effects, restricting ourselves to "normal" shrinkage not associated with collapse.

Wood is usually green when it is cut from the log and must be dried for most uses to a moisture content suitable for use conditions. Figure 3.20 shows the ef-

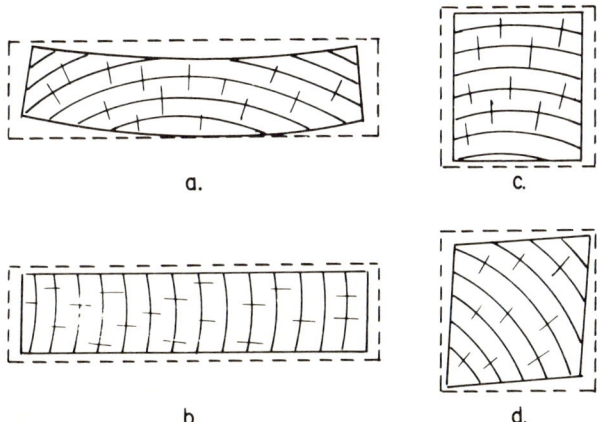

Fig. 3.20. Change in cross-sectional shapes after drying of *a.* flat-sawn board, *b.* quarter-sawn board, *c.* and *d.* square sections.

fect of the anisotropic transverse shrinkage on the change in shape of lumber after drying.

The most striking change is in the flat-sawn board *a* which cups into a trough-like shape after drying. This effect can be reduced if the lumber is restrained from cupping during drying by the application of weights or other restraints to force the lumber to remain more nearly flat. However cupping cannot be eliminated completely since some residual stress which tends to cup the board is inevitably present. If a plane board is required for use, considerable surfacing of the cupped surfaces is required, resulting in a thinner board and much waste. Furthermore, after exposure to changing environmental humidity the board will tend to cup cyclically with cyclic humidity changes.

The quarter-sawn board **b** does not cup, although its shape changes somewhat as shown. This is the best way of cutting lumber from the viewpoint of dimensional stability but in the normal practice of sawing a log it is only possible to obtain one or two truly quarter-sawn boards, the others having a ring orientation somewhere between the two cases shown in **a** and **b**.

The square timber shown in **c** becomes somewhat rectangular after drying. That shown in **d** becomes diamond-shaped and requires considerable surfacing to obtain right-angled edges. It is clear, therefore, that it is better to cut square timbers in the manner shown in **c**. Again, however, in order to obtain maximum lumber from the log, it is necessary to cut material which is like **d** or between **c** and **d**.

A number of theories have been proposed to explain transverse shrinkage anisotropy. These have been reviewed by Pentoney (1953), Bosshard (1956), Kelsey (1963), Stamm (1964), Crews (1965), Kollmann and Côté (1968), and others. Pentoney has divided these theories into three groups as follows, based on consideration of gross wood structure (arrangement of cell types), fibril alignment modifications, and cell-wall layering variations.

There are two theories for explaining transverse anisotropy based on gross wood structure. The first is the ray restraint theory, based on the assumption that ray tissue shrinks less radially than does longitudinal tissue and therefore restrains radial shrinkage. The second is the earlywood-latewood interaction theory which postulates that the latewood shrinks more tangentially than the earlywood, forcing the latter to shrink more tangentially and less radially and thus causing transverse shrinkage anisotropy. Both of these theories are discussed in detail below.

It is more convenient to discuss differential radial and tangential shrinkages in terms of the coefficients X_r and X_t as defined in equations (3.34) and (3.35) rather than in terms of total shrinkages s_r and s_t. The coefficients X_r and X_t are positive for swelling or shrinking since in the latter case the radial r or tangential t dimensions decrease as moisture content decreases and therefore dr/dm and dt/dm are positive. At a given moisture content the swelling and shrinkage coefficients are equal provided the phenomena are reversible. We will assume so for purposes of the present analysis.

The *ray restraint* theory can best be understood by reference to Figure 3.21 which shows a hypothetical cross-section for wood consisting of longitudinal and ray tissues. The assumption is made that the radial shrinkage coefficient X_{rr} of isolated ray tissue is lower than the radial shrinkage coefficient X_{r0} of ray-free longitudinal tissue. The radial shrinkage coefficient X_r of the gross wood including rays should be intermediate between X_{rr} and X_{r0}, the actual value depending on the relative stiffness of the ray and ray-free tissues. The relative stiffness K_r of the ray tissue is given by

$$K_r = (E_r V_r)/(E_r V_r + E_0 V_0) \qquad (3.42)$$

where E_r and E_0 are the Young's moduli and V_r and V_0 are the volume fractions of the ray and ray-free tissues. The relative stiffness K_0 of the ray-free tissue is

$$K_0 = 1 - K_r = (E_0 V_0)/(E_r V_r + E_0 V_0). \qquad (3.43)$$

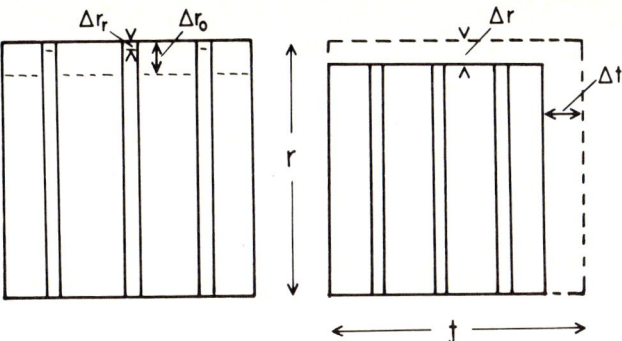

Fig. 3.21. Schematic diagram of model used for the ray-restraint theory of anisotropic transverse shrinkage showing a. radial shrinkages of isolated ray tissue Δr_r and of ray-free tissue Δr_0, and b. radial Δr and tangential Δt shrinkages of the gross wood.

The anticipated value of X_r can be calculated approximately from

$$X_r \approx X_{rr}K_r + X_{r0}K_0 \approx (X_{rr}E_rV_r + X_{r0}E_0V_0)/(E_rV_r + E_0V_0) \quad (3.44)$$

based on the assumption that the strains are elastic or at least proportional to the stresses. Equation (3.44) can be rearranged and written in the form

$$X_r \approx \frac{X_{rr}V_r + X_{r0}(1 - V_r)(E_0/E_r)}{V_r + (1 - V_r)(E_0/E_r)} \quad (3.45)$$

Equation (3.45) is an approximation to a more exact equation given by McIntosh (1955). It differs also in that his equation is given in terms of percent radial shrinkage s_r of the wood in terms of the percent radial shrinkage s_{rr} of the ray tissue and s_{r0} of the ray-free tissue. McIntosh measured the isolated shrinkages s_r of ray tissue from red oak and American beech wood, and of wood containing various volume fractions V_r of ray tissue, using microtomed sections. He then measured the radial shrinkage s_{rr} of isolated rays and the radial shrinkage s_r of wood containing two different fractions V_r of ray tissue. From these two sets of measurements he was able to calculate both s_{r0} and the ratio E_0/E_r by solving his equation for two different sets of measured data on the other variables s_r, V_r and s_{r0}, the latter value being taken as constant. The results he obtained for beech are shown in Figure 3.22.

It is informative to note how the calculated values of radial wood shrinkage without rays compares with the observed radial shrinkage of the gross wood and of the rays themselves as listed in Table 3.5. Also shown are data obtained from several other sources which give the radial shrinkage of ray tissues and also of rayless wood and of wood containing rays. The data clearly indicate that the ray restraint mechanism is at least one of the factors responsible for transverse shrinkage anisotropy.

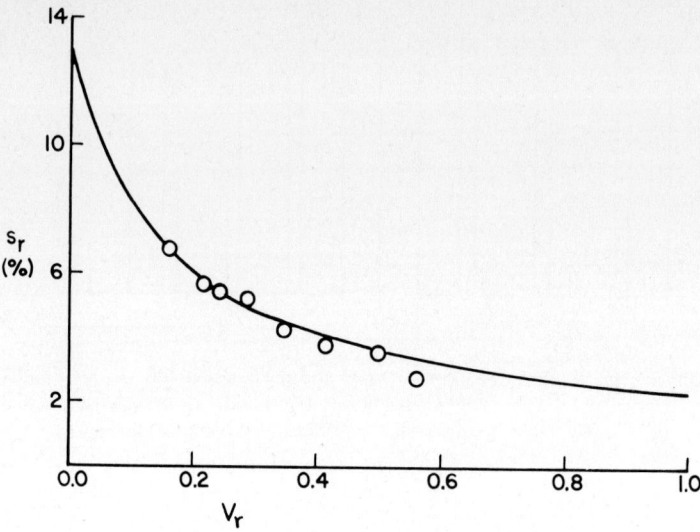

Fig. 3.22. Experimental points and calculated curve of percent radial shrinkage s_r against fraction of ray tissue V_r. The curve is calculated based on an E_0/E_r ratio of 0.124 (adapted from McIntosh 1955).

Table 3.5. Percent Radial Shrinkage from Wet to Dry Condition of Wood Rays, Rayless Wood, and Wood Containing Rays (taken from various sources)

Reference	Kind of Wood	Radial Shrinkage Wet to Dry (%)			
		s_{rr}*	s_{r0} †	s'_{r0} ‡	s_r §
Greenhill (1944)	*Casuarina luckmanni*	1.2	—	—	3.3
	Grevillea robusta	1.2	—	—	3.7
	Xylomelum pyriforme	0.7	—	—	2.0
	Quercus sp	3.2	—	—	4.9
Lindsay and Chalk (1954)	*Quercus ilex*	3.1	—	6.0	—
	Cardwellia sublimis	0.9	—	3.5	—
	Helicia terminalis	0.8	—	1.9	—
Morschauser and Preston (1954)	*Quercus borealis*	2.5	12.0	—	5.1
McIntosh (1955)	*Quercus borealis*	2.6	6.8	4.8	—
	Fagus grandifolia	2.3	12.7	6.7	—
de S. Wijesinghe (1959)	*Arctocarpus integra*	1.3	4.8	—	3.8
	Canarium zeylanicum	0.2	4.0	—	3.4
	Acer pseudoplatanus	3.8	4.9	—	4.3
Schniewind and Kersavage (1962)	*Quercus kelloggii*	2.1	—	5.8	3.0

*s_{rr} radial shrinkage of isolated rays.
†s_{r0} radial shrinkage of ray-free wood.
‡s'_{r0} radial shrinkage of wood containing fine rays but no broad rays.
§s_r radial shrinkage of gross wood including all rays.

The ray restraint theory has been criticized by Ritter (1939) because his X-ray and microscopic measurements of the ray cell-wall fibril angles of basswood indicate that they have essentially the same orientation with respect to the longitudinal axis of the gross wood as do the longitudinal cells. Therefore they should shrink nearly the same amount in the radial direction as do the longitudinal tissues. In view of the fact that the ray tissues for many woods appear to shrink less radially than do the longitudinal tissues, it is difficult to resolve these findings. It is possible that basswood is different from most woods with respect to the fibril orientation in the rays. According to Crews (1965), "the shrinkage behavior of basswood is the exception rather than the rule," since it exhibited very different shrinkage patterns than the other four hardwoods which he studied. Studies by Harada and Wardrop (1960) showed that the fibril alignment in the $S2$ layer of ray parenchyma cells in *Cryptomeria japonica* is essentially parallel to the cell axes, or radially oriented in the tree. This would support the concept that rays shrink less radially than longitudinal tissues.

Another possibility is that the hygroscopicity of ray tissue is lower than that of longitudinal tissue which would reduce the total shrinkage. For example, Morschauser and Preston (1954) report that the fiber-saturation point M_f was only 17.63 percent for ray tissue, compared with 25.57 percent for longitudinal tissue for *Quercus borealis*, based on radial-shrinkage intersection measurements. They also report that the ray tissue was less dense than the longitudinal tissue, which they speculated might also affect the relative radial shrinkages of the two tissues.

In the above discussion no mention was made of tangential shrinkage or of any possible coupling of tangential and radial or longitudinal shrinkages through Poisson ratio* effects. Any reduction in radial shrinkage of the longitudinal tissue caused by ray restraint would tend to increase tangential shrinkage because of Poisson ratio effects in longitudinal tissue. This may be partially reduced by the same effect in the ray itself in that increased radial shrinkage in the ray would be expected to produce decreased tangential shrinkage in the ray. However, this effect should be smaller than the increase in tangential ray shrinkage caused by the great reduction in ray shrinkage parallel to the axes of the longitudinal cells. This is true for two reasons. First, the amount of restraint of the rays parallel to the grain of the wood may be greater than that in the radial direction since there is essentially no shrinkage of the ray permitted parallel to the grain because of the low longitudinal shrinkage of the wood. Second, the Poisson ratio would be expected to be higher in the longitudinal-tangential plane than in the radial-tangential plane because of the orientation of the cell axes in the radial direction. A more detailed analysis of the Poisson ratio effect and its influence

---

*Poisson's ratio is the ratio of the strain in one direction of a solid body to the strain in a second direction, perpendicular to the first direction, caused by a stress applied in the second direction.

on anisotropic transverse shrinkage is given in later paragraphs along with discussion of the earlywood-latewood interaction theory.

Crews (1965) found that ray shape was a factor in influencing the T/R shrinkage ratio in five United States hardwoods. Short wide rays reduced the radial
shrinkage more than high narrow rays. Crews attributed the relative ineffective-
ness of narrow rays to their low stiffness in the radial direction. He also concluded that the effective radial shrinkage restraint of ray tissue decreases as the
over-all specific gravity of the wood increases. This is in agreement with equation (3.45) since the terms E_0 and X_{r0} would then increase as the specific gravity of the wood increases for a given ray volume, giving an increase in the radial
coefficient X_r.

The *earlywood-latewood interaction* theory for explaining transverse shrinkage
anisotropy has been called the Mörath theory by Pentoney (1953) because
Kollmann (1936) attributed it to Mörath. However, Erickson (1955) points out
that the general idea behind the theory has been known at least as far back as
Wagner (1917) and Forsaith (1926).

This theory (Figure 3.23) states that the tangential shrinkage is greater than

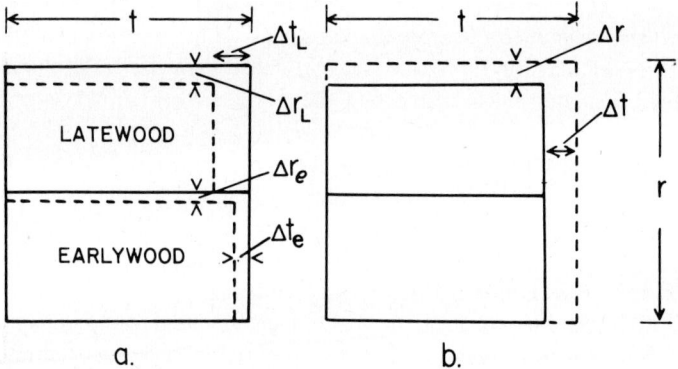

Fig. 3.23. Schematic diagram of model used for earlywood-latewood theory of anisotropic transverse shrinkage showing *a*. radial and tangential shrinkages of isolated earlywood
(Δr_e, Δt_e) and of isolated latewood (Δr_l, Δt_l), and *b*. radial Δr and tangential Δt shrinkages
of the gross wood.

radial shrinkage because of the alternation of earlywood and latewood in the radial direction in many woods grown in the temperate zone. The shrinkage of
latewood is greater than that of earlywood, and the latewood is much stronger
than the earlywood. Therefore, the strong tangential bands of latewood force
the weak bands of earlywood to shrink tangentially to about the same extent as
the latewood. In the radial direction, however, the total shrinkage is nearly equal
to the weighted mean shrinkage of the two components. Actually, it is less than
this because of the increase in radial dimension of the earlywood caused by ex-

cessive shrinkage in the tangential direction through the Poisson effect. There is also some increase in the radial shrinkage of the latewood because its tangential shrinkage is somewhat reduced by the resistance of the earlywood to excessive tangential shrinkage.

For the expected tangential shrinkage coefficient X_t based on earlywood-latewood interaction an approximate equation similar to equation 3.45 and with similar assumptions can be derived. It can be written

$$X_t \approx \frac{X_{te}V_e + X_{tl}V_lE_l/E_e}{V_e + V_lE_l/E_e} \tag{3.46}$$

where V_e, E_e, and X_{te} are the volume fraction, tangential Young's modulus, and tangential shrinkage coefficient, respectively, of the isolated earlywood, and V_l, E_l, and X_{tl} are the corresponding terms for the latewood.

An approximate expression can also be written for the expected radial-shrinkage coefficient X_r. This expression is more complex because of the Poisson's ratio effect. If it is assumed that the Poisson's ratio $\cup$ is the same for latewood and earlywood, then the expansion for X_r can be written

$$X_r \approx \frac{(X_{re}V_e + X_{rl}V_l)[V_e + V_l(E_l/E_e)] + V_e V_l[\cup(X_{tl} - X_{te})(1 - E_l/E_e)]}{V_e + (V_lE_l/E_e)} \tag{3.47}$$

where the additional terms X_{tl} and X_{te} are the tangential shrinkage coefficients of the isolated latewood and earlywood. The anticipated ratio X_t/X_r, often designated as the T/R ratio, can be obtained by dividing equation (3.47) by (3.46). It will not be given here since it is merely the ratios of the numerators of the equations, the denominators being identical.

Pentoney (1953) gives a more exact equation for the anticipated T/R shrinkage ratio in terms of the total directional shrinkages s_{re}, s_{rl}, s_{te}, and s_{tl} of isolated earlywood and latewood, rather than the shrinkage coefficients X. These equations in terms of T/R ratios are essentially similar to those given above if it is assumed that

$$100 - s_{te} \approx 100 - s_{tl} \tag{3.48a}$$

$$100 - s_{re} \approx 100 - s_{rl} \tag{3.48b}$$

where the shrinkages are given in percent.

Ylinen and Jumppanen (1967) developed and tested a more rigorous equation for testing the earlywood-latewood interaction theory. They also extended the theory into longitudinal shrinkage on the assumption that such shrinkage is greater for earlywood than for latewood, as is usually observed.

Pentoney (1953) tested the earlywood-latewood interaction hypothesis experimentally by measuring the tangential and radial shrinkages of isolated early-

Table 3.6. Average Shrinkage of Douglas-fir Latewood and Earlywood from Fiber
Saturation to the Ovendry Condition (Pentoney 1953)

Test Condition	Tangential Shrinkage (%)		Radial Shrinkage (%)	
	Latewood	Earlywood	Latewood	Earlywood
Isolated	$7.21 \pm .93$	$4.81 \pm .80$	$8.90 \pm .99$	2.39 ± 1.05
Attached	$7.53 \pm .60$	$7.06 \pm .71$	7.76 ± 1.68	1.36 ± 1.41

wood and latewood of Douglas-fir, and compared these with those obtained in the gross wood with the two kinds of tissues attached. The results are shown in Table 3.6. It is clear that the tangential shrinkage of the latewood is essentially the same whether isolated or attached to earlywood. The earlywood when attached to latewood has been forced to shrink essentially the same tangentially as the strong latewood. It also appears, as the theory suggests, that the radial shrinkage of the earlywood is decreased when it is attached to the latewood. The isolated earlywood shrinks much less than the isolated latewood in both directions, as anticipated. Furthermore, the isolated earlywood shows a T/R ratio of close to 2, while the isolated latewood T/R ratio is slightly less than unity. The high T/R ratio for the earlywood indicates that some mechanism other than the earlywood-latewood interaction may be operating in the earlywood.

Pentoney has shown that it is possible to calculate hypothetical T/R shrinkage ratios for woods of given specific gravity ratios G_l/G_e of latewood to earlywood and of different T/R ratios within isolated earlywood and latewood by use of equations similar to equations (3.46) and (3.47). For example, if G_l/G_e is 2.5 and it is assumed that $X_{re} = X_{te}$ and $X_{rl} = X_{tl}$, that $\cup = 0.5$, and that $X_{rl} = 2.5 X_{re}$ ($X_{tl} = 2.5 X_{te}$), the expected values of X_t and X_r and the T/R (X_t/X_r) ratio can be calculated as functions of the fraction V_e of earlywood. It is known that E_l/E_e varies as $(G_l/G_e)^{2.5}$ (*Wood Handbook*) or since $(2.5)^{2.5} \approx 10$, $E_l/E_e \approx 10$. Figure 3.24 shows curves of X_r/X_{re} ($= X_r/X_{te}$), of X_t/X_{te} ($= X_t/X_{re}$), and of the T/R ratio (X_t/X_r) calculated from equations (3.46) and (3.47).

Earlier work by Vintila (1939) on four softwoods tends to support the earlywood-latewood interaction theory for anisotropic transverse shrinkage, as do those by Ritter and Mitchell (1939), Browne (1957), and Erickson (1955). Their data are tabulated in Table 3.7. It is interesting to note that Vintila found large T/R shrinkage ratios in isolated earlywood compared with those for isolated latewood for all four woods, in agreement with the results of Pentoney. This follows the general observation that the T/R shrinkage ratio decreases with increasing wood density.

Browne's (1957) data on Douglas-fir and southern yellow pine show T/R shrinkage ratios for isolated latewood which are much lower than unity, as does the data of Pentoney on Douglas-fir. This finding is difficult to resolve in terms of the presently described mechanisms. Browne also pointed out an interesting

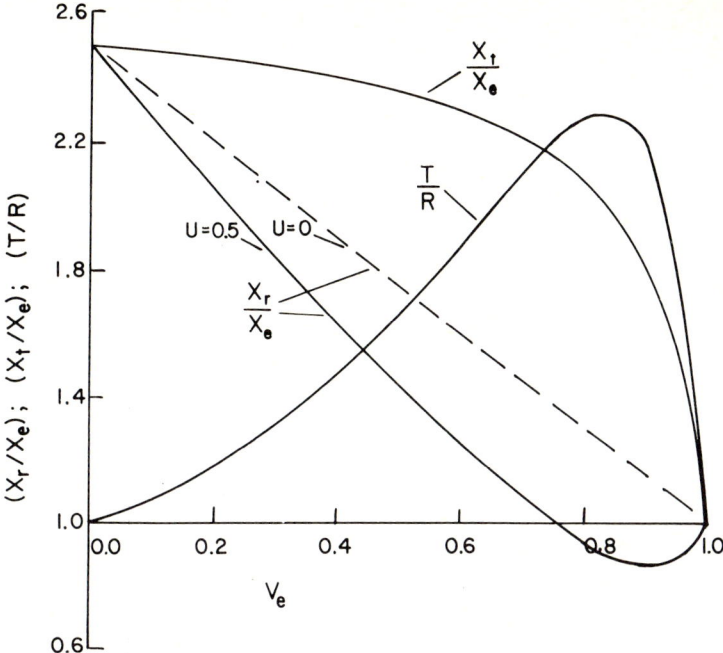

Fig. 3.24. Calculated curves of X_t/X_e, X_r/X_e, and X_t/X_r (T/R ratio) against fraction of earlywood, assuming a Poisson's ratio $\cup = 0.5$. Also shown is X_r/X_e for $\cup = 0.0$.

Table 3.7. Fractional Shrinkage of Isolated Earlywood and Latewood Tissue in Tangential and Radial Directions, and T/R Shrinkage Ratios

Reference	Kind of Wood	Earlywood			Latewood		
		s_{et}	s_{er}	T/R	s_{lt}	s_{lr}	T/R
Vintila	Douglas-fir	.057	.029	2.0	.109	.098	1.1
(1939)	Pine	.080	.029	2.8	.113	.082	1.4
	Larch	.071	.032	2.2	.122	.102	1.2
	Fir	.050	.024	2.1	.088	.062	1.4
Pentoney	Douglas-fir	.050	.024	2.1	.078	.098	0.8
(1953)							
Browne*	Douglas-fir	.067	.056	1.2	.086	.123	0.7
(1957)	S. Yellow pine	.064	.077	0.8	.073	.130	0.6
Ritter and Mitchell	Longleaf pine	.066	—	—	.093	—	—
(1939)							
Erickson	Douglas-fir	.043	—	—	.093	—	—
(1955)	W. redcedar	.039	—	—	.074	—	—

*Data are fractional swelling from dry to water-soaked condition. All others are shrinkage figures from water-soaked to dry condition.

optical illusion which may lead to errors in measuring *in-situ* shrinkage on woods such as southern yellow pine which exhibit pronounced color differences between earlywood and latewood. He found that the edges of the latewood bands in dry flat-sawn boards appeared to migrate into the earlywood when the boards were soaked in water. He described this effect as being the result of the transparency of surface cells of water-soaked earlywood which caused cells of latewood deeper in the wood to appear to be on the surface, and to give the effect of a reduced earlywood and an excessively increased latewood width on the water-soaked flat-sawn boards.

The effect of fibril angle on the longitudinal-transverse shrinkage anisotropy of wood has already been demonstrated. It has been proposed by Ritter and Mitchell (1939) and by Frey-Wyssling (1940) that fibril angles are greater in the radial than in the tangential walls, the difference ranging up to 15°. Ritter and Mitchell attribute the greater fibril angle in the radial walls largely to the circular orientation of the crystallites or microfibrils around the pits which are so abundant on these walls, in softwoods particularly.

Pentoney (1953) has calculated that a small fibril angle difference of 15° cannot account for the large T/R shrinkage ratio observed for wood, based on the lattice model proposed by Cockrell (1947). According to Pentoney the radial or tangential shrinkage should be proportional to the cosine of the fibril angle. Pentoney calculated that if the mean fibril angle θ in the radial walls was 30° and in the tangential walls 15°, the T/R ratio should be equal to (cos 15°)/ (cos 30°), or 1.11, which is much smaller than the commonly observed ratio of 2. However, based on the simplified model of Barber and Meylan (1964) discussed earlier, there should also be a greater reduced shrinkage in the thickness of the tangential wall than of the radial wall which would tend to increase the T/R ratio to a still greater extent. This is qualitatively clear from Figure 3.16*d* which shows that the decrease in transverse shrinkage (or swelling) in going from a fibril angle θ of 15° (tangential wall) to one of 30° (radial wall) is accompanied by an increase in the thickness of the corresponding wall. The figure also shows the relative dimensional changes in the two walls as anticipated from the model. Barber and Meylan (1964) should be consulted for a more quantitative explanation of this effect.

The correlation between the anisotropy of shrinkage and of dielectric properties in isolated earlywood and latewood, observed by Nakato (1958), led him to conclude that differences in the orientation of active hydroxyl groups in the radial and tangential walls was responsible for both phenomena. This may therefore be classified as a theory based essentially on variations in the fibril angle. It is known that there are large differences in the dielectric properties of wood measured with the electric field parallel to the grain and in the transverse direction, and it has been postulated that they are at least partially caused by the difference in fibril orientation in the two directions (Skaar 1948).

The third category of theories for explaining transverse-shrinkage anisotropy is based on differences in the swelling characteristics and arrangements of the various cell wall layers.

Frey-Wyssling (1940b) has proposed that variations in the compound middle lamella cause the transverse-shrinkage anisotropy in wood. He found, for example, that there are more crosswalls in larch (and therefore more layers of middle lamella) per unit length in transverse microtome sections in the tangential than in the radial direction (Figure 3.25a). Furthermore, he also found (Figure 3.25b) that the thicknesses of the individual middle lamellae were greater in the

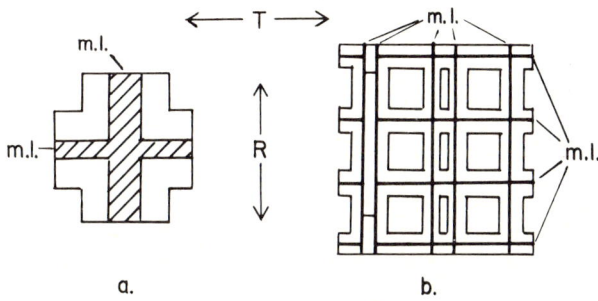

a.   b.

Fig. 3.25. Model used for explaining anisotropic transverse shrinkage based on the middle lamella variations in radial and tangential directions showing a. thicker middle lamella in tangential direction, and b. greater number of middle lamella layers in the tangential direction (adapted from Pentoney 1953).

radial than in the tangential walls. He proposed that there is more thickness shrinkage in the middle lamella than in the secondary wall and therefore more shrinkage in the tangential than in the radial direction. The high concentration of hygroexpansive pectin substances in the middle lamella region may cause greater swelling in these layers. This theory has not been confirmed, and Matsumoto (1954), according to Crews (1965), has concluded instead that little shrinkage occurs in the middle lamella and that most shrinkage is confined to the secondary walls.

Bosshard (1963) suggests that there is a higher lignin content in radial than in tangential walls of longitudinal wood cells. This therefore restricts radial shrinkage somewhat since the lignin is stiffer and less hygroscopic than the holocellulosic fractions of the cell wall.

It seems apparent from the above discussions that there may be several mechanisms which operate to cause transverse anisotropy in wood shrinkage and that these are not necessarily identical for all woods. Furthermore, considerable controversy exists concerning the relative importance of the various mechanisms which have been proposed.

There is evidence that the T/R shrinkage ratio is not constant over the entire hygroscopic moisture range. Keylwerth (1962), for example, measured the tangential and radial swelling of European beech from the dry condition to various moisture contents and found that the ratios of the two directional swellings increased with moisture content as shown in Figure 3.26. Noack (1964) has found

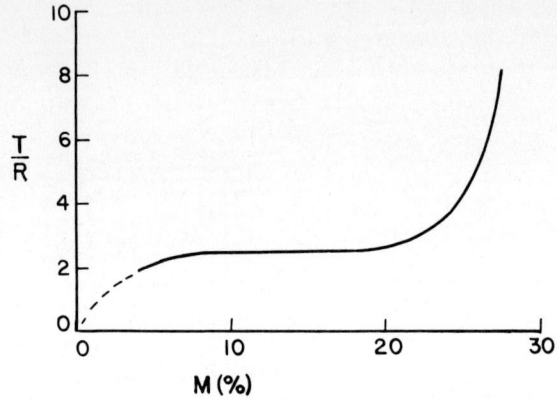

Fig. 3.26. Curve of X_t/X_r (T/R ratio) against wood moisture content M (%) for European beech (adapted from Keylwerth 1962).

somewhat similar results, as has Hittmeier (1967) working with Engelmann spruce. He suggests that radial swelling is complete at moisture contents 3–5 percent below that at which tangential swelling ceases. Furthermore, this would agree with the results of Kelsey (1956) who found that the shrinkage intersection point, that is the moisture content below which hygroscopic shrinkage begins, is lower for radial than for tangential shrinkage. The reason for this apparent anomaly is not known.

Swelling beyond the completely water-swollen dimensions in concentrated aquaeous salt solutions is almost entirely tangential (Stamm 1964). Some woods swell more radially than tangentially in liquid ammonia. (Stamm 1955; Pollisco, Skaar, and Davidson 1971; and R. A. Parham, R. W. Davidson, and C. H. de Zeeuw 1972.)

Shrinkage and Stress

The shrinkage and subsequent hygro-dimensional changes in wood are affected to a greater or lesser degree by mechanical stresses. These stresses may arise in several ways including moisture gradients, mechanical restraints, macroscopic tissue swelling anisotropy, and microscopic and submicroscopic anisotropy within the cell wall itself. In any of these cases the effect on dimensional change may be indirect in that EMC itself is changed because of the Barkas effect, discussed in Chapter 4, or it may be a direct mechanical deformation either elastic or inelastic or both. It is impossible in practice completely to eliminate stresses during moisture changes in wood, particularly those associated with the third and fourth items mentioned above. We will discuss each of these types of stresses and their effects on dimensional changes in the order given above.

Moisture gradients always exist in wood under normal conditions, not only during initial drying but also during subsequent use, since atmospheric humidity and temperature are subjected to both cyclic and random changes. The stresses associated with drying of lumber have been studied in some detail and actually form the basis for most lumber-drying schedules. Figure 3.27, taken from

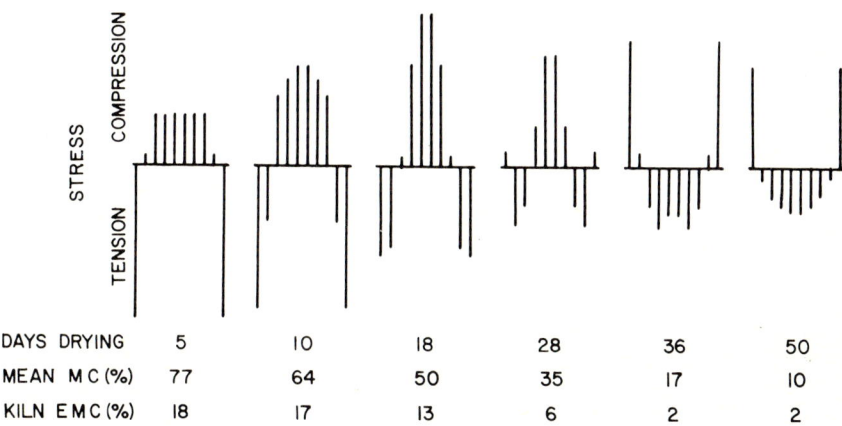

DAYS DRYING	5	10	18	28	36	50
MEAN M C (%)	77	64	50	35	17	10
KILN E M C (%)	18	17	13	6	2	2

Fig. 3.27. Residual stresses at various stages of drying (adapted from McMillen 1958).

McMillen (1958), shows a typical sequence of transverse residual stresses in lumber during normal drying schedules. The phenomenon of "casehardening," for example, is associated with the high transverse-tensile stress which exists during the early stages of drying near the surface of the lumber. It is instructive to follow the sequence of moisture gradients and associated stresses during a normal drying schedule and to interpret these in terms of their effect on initial drying shrinkage.

During the initial stages of normal drying the surface dries below fiber saturation and some hygroscopic shrinkage takes place. However, this is resisted by the wet core which is above saturation. As a consequence, the situation shown in Figure 3.27 after five days of drying occurs. Here the outer layers of the wood are under severe tensile stress whereas the inner layers are under a milder compressive stress. Because of the long time duration of drying and the strong tendency of wood to "creep" under conditions of stress, particularly transverse to the grain, the outer layers become "stretched" or "set" in tension and shrink less than they would if drying gradients did not exist. Often the tensile stress which occurs during this early stage of drying may cause checks or small cracks to appear on the surface of the lumber. These normally close and are not as visible during later stages of drying, but they are an undesirable feature in subsequent use of the wood, particularly if the surface is to be exposed to cyclic humidity conditions. Furthermore, they cause problems if the surface is to be coated with paint or clear finishes, since these penetrate into the checks and make them visible, particularly with clear finishes.

As drying progresses further, as shown for the ten- and eighteen-day drying times in Figure 3.27, the compressive stress in the interior increases as intermediate layers dry and commence to shrink. The tensile stress in the extreme outer layer decreases, however, because the adjoining layers begin to develop some tensile stress as they try to shrink relative to the inner undried layers. After twenty-eight days the outer layer is under compression and it remains so through the subsequent drying time. This occurs because the outer layer has been stretched or set severely in tension during the initial drying period compared with adjoining layers. Therefore, as the intermediate layers dry below fiber saturation they tend to shrink to smaller dimensions parallel to the wide face of the board than the adjoining surface layers, even though the latter may be somewhat drier. Finally, when the wood is nearly uniformly dry after fifty days the stresses have essentially reversed themselves from those existing after five days, as Figure 3.27 indicates. This final distribution of stresses is that which is characteristic of so-called casehardened lumber.

The term "casehardening" as applied to dry lumber does not indicate that the surface is mechanically any harder than wood which is not casehardened. The term is really a misnomer but has probably been adapted from the field of metallurgy because of the analogy between stresses arising from moisture gradients during drying of lumber and stresses caused by temperature gradients near the surfaces of metal objects when they are casehardened as a part of the manufacturing process in order to harden the outer surface.

Casehardening is not a serious problem in lumber unless it is resawn into thinner pieces in which case the residual stresses cause excessive cupping of the surfaces. For ordinary construction lumber, for example, it has little importance. Casehardening is removed, if necessary, by subjecting the lumber to air of high humidity and temperature for a short time near the end of the drying process. This causes additional compressive stress in the outer surface sufficient to cause the reverse kind of creep which brought about the original casehardening and if properly carried out results in stress-free lumber. If casehardening is overcorrected by application of too high a humidity-temperature combination for too long a time period, the wood may attain the condition known as reverse casehardening. As the name implies, this is a state in which the reverse stresses exist in the dry lumber. In other words, the surface layers now have a compression set and are therefore in a state of tension compared with the interior layers of wood when the wood is uniformly dry.

The normal test for casehardening is to saw out a section of the interior of the wood from a cross-section of a dry board of $\frac{1}{2}$ to 1 inch thickness. Figure 3.28,

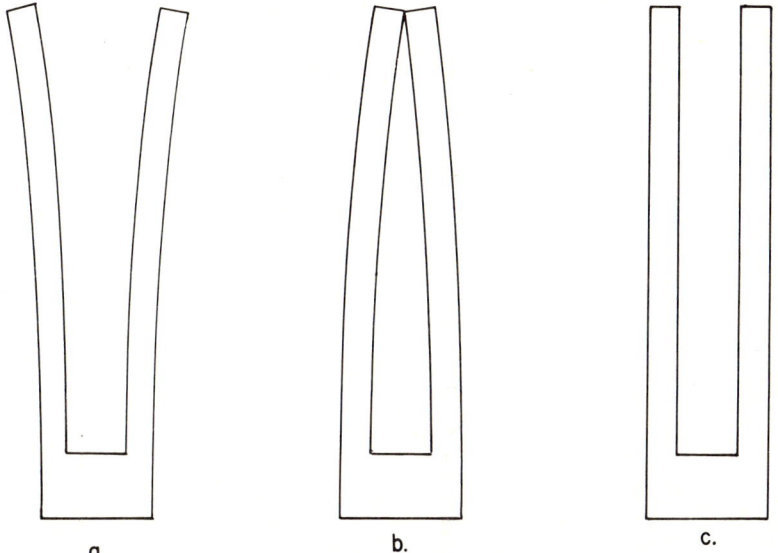

a.   b.   c.

Fig. 3.28. Typical appearance during casehardening tests of the cross-section of wood samples with *a.* reverse casehardening, *b.* casehardening, and *c.* no residual drying stresses (adapted from McMillen 1958).

taken from McMillen (1958), shows the shapes which are attained by the section in wood with casehardening in the stress-free condition and with reverse casehardening when the wood is uniformly dry.

It is clear that the outer surfaces of normally dried wood containing casehardening have a tension set and are in a stretched condition. There thus tends to be less shrinkage across the width of the board in these outer layers than for wood dried without stress. However, the inner layers have been somewhat compressed during the early stages of drying and therefore tend to show greater shrinkage than unstressed wood. There is much more volume in the interior with compression set than exists in the surface layers with tension set. Therefore, although the tension set in the surface layers is usually more pronounced than the compression set, the over-all shrinkage across the width of the board is determined largely by the amount of compression set. According to McMillen (1958), the amount of compression set in the board interior, and therefore the extent of shrinkage, is minimized by use of low temperatures and humidities during the drying operation.

Espenas (1971) made a study of the comparative transverse shrinkage of Douglas-fir, western hemlock, and red alder when dried from the green condition at temperatures of 90, 150, 180, 200, and 215°F at relative humidities designed to give nominal EMC's of 6, 9, and 12 percent. His results confirmed the fact that greater shrinkage occurs during the drying at higher temperatures. The difference was greatest for the samples dried at 12 percent EMC conditions. The increase in tangential and radial shrinkages in material dried at 215°F compared with that dried at 90°F, for example, was 33 and 51 percent for Douglas fir, 65 and 82 percent for western hemlock, and 130 and 137 percent for red alder. Only a small portion of these differences could be attributed to the decrease in hygroscopicity of the wood associated with drying at the higher temperatures (see Chapter 2).

Additional evidence to support the effect of stresses arising from moisture gradients in wood is shown in Figure 3.29, taken from Hittmeier (1967). This shows curves of the squares of fractional radial and tangential swelling (the ratio of the swelling at time t to the total equilibrium swelling) as functions of time

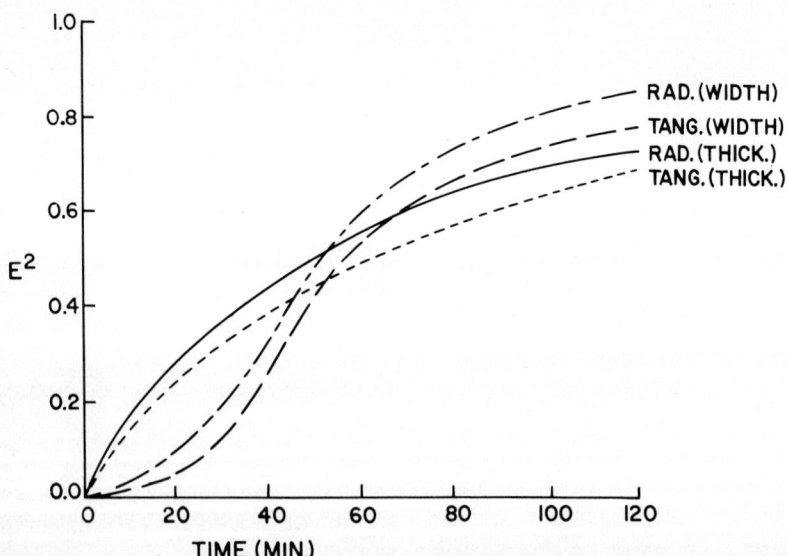

Fig. 3.29. Curves of fractional radial and tangential swelling squared (E^2) against time in the width and thickness direction for quarter-sawn (radial width and tangential thickness) and flat-sawn (tangential width and radial thickness) samples showing the delayed swelling in width in both cases (adapted from Hittmeier 1967).

for initially dry wood during soaking in distilled water at 20°C. Two of the curves are for flat-sawn samples 0.3 cm thick radially by 3 X 3 cm tangentially and longitudinally, and two curves are for quarter-sawn samples 0.3 cm thick tangentially by 3 X 3 cm radially and longitudinally.

It is apparent from Figure 3.29 that the width swelling is delayed compared with the thickness swelling for both flat-sawn and quarter-sawn samples. According to Hittmeier this is attributed to the fact that most of the diffusion of water into the dry sample takes place in the direction of thickness. During the early stages of soaking, outer layers take up water and tend to swell. They are restrained from swelling in the width direction, however, because the dry interior wood is stronger than the moist surface layers. Swelling does take place in the thickness direction and may even be somewhat greater than normal because of the Poisson ratio effect in the moist surface layers. Later, as the interior wood finally becomes moist, the width swelling increases more rapidly than the increase in thickness. The stress pattern in the wood is nearly the reverse of that which occurs during the normal drying of lumber. Hittmeier carried out similar experiments with cross-sectional samples. In this case there was little restraining effect because the water penetrated the cell cavities in the center of the samples almost immediately and the rate of penetration of the cell wall was essentially the same throughout the sample thickness.

The phenomenon described above is very similar to that observed by Hermans (1946) who measured the length and diameter changes with time of soaking in water of model cellulose fibers which were initially dry. He noted that the diameter (analogous to thickness in Hittmeier's experiments) increased with time immediately after immersion of the fiber in water. However, the length (analogous to width of sample in Hittmeier's experiments) remained essentially unchanged for ten minutes until some water penetrated to the center of the fiber. Thereafter the length swelling increased more rapidly than the diameter swelling. Hermans was also able to study the penetration of the water boundary into his fiber sample and noted that the increase in length began at nearly the same time that the boundary penetrated to the center of the fibers.

Sadoh and Christensen (1967), in their study of longitudinal shrinkage and swelling of *Araucaria cunninghamii*, noted a hysteresis effect. They attributed this in part to the stresses associated with drying even in thin microtome sections.

The effect of mechanical restraint on hygroscopic dimensional changes is very marked in wood because of the large component of rheological or inelastic deformation associated with mechanical stress. This of course is the cause of casehardening as described previously. The inelastic deformation under mechanical restraint can be advantageous or not depending on the circumstances. For example, mechanical restraint is used to reduce cupping, twisting, bowing, and crooking of lumber caused by diagonal grain or anisotropic shrinkage during drying (Rasmussen 1961). Koch (1971) has used similar techniques to produce straight 2 X 4 inch lumber from juvenile wood of southern yellow pine which is prone to crook, bow, and twist due largely to excessive longitudinal shrinkage.

The deleterious effects of mechanical stress on shrinkage are most apparent when wood which is restrained from movement is subjected to cyclic moisture changes. This effect shows up, for example, in wood flooring which is used indoors in temperate climates where marked annual changes in indoor humidity occur (see Chapter 2). If the flooring is fitted tightly during installation at a moisture content lower than the highest which it attains during the subsequent humidity cycles, then it will be subjected to a compressive transverse stress during the high humidity portion of the cycle. This tends to buckle the flooring, but if the flooring is adequately fastened the wood relaxes and develops a compression set similar to that described in the previous section. The transverse dimension of each board is now smaller at any given moisture content than when it was installed. Therefore, the spaces develop between the boards and the appearance of the floor deteriorates.

The loosening of wood fittings subjected to cyclic moisture changes is also accelerated by the development of compression set due to restrained swelling. One of the more striking examples of the effect of restrained swelling on subsequent wood dimensions is found in the loosening of wooden axe handles and other tool handles which are soaked in water to tighten them. They are tightened as the wood handle swells against the restraint imposed by the tool head but in time the swelling pressure relaxes causing an excessive compression set. Thus the portion of the handle inserted into the hammers or axe head now attains a smaller dimension than the original dimensions when it is exposed to normal atmospheric conditions again. Figure 3.30 shows the sequence of dimensional changes in the wood during a cycle as described above.

Tiemann (1944) has demonstrated a similar effect in which the wood is restrained from shrinkage in one dimension during drying. In this case the wood is "stretched" in the direction of restraint having attained a tension set and is of larger dimensions in this direction than the unrestrained wood.

The effect of macroscopic tissue shrinkage anisotropy, and the development of mechanical stresses thereby, has already been discussed in connection with the theories of anisotropic shrinkage. Both the ray restraint and the earlywood-latewood interaction theories are based on the anisotropy of shrinkage of the respective tissues and the mechanical stresses which arise from this differential shrinkage.

The microscopic and submicroscopic anisotropy of shrinkage within the cell wall itself has also been discussed previously in this chapter. It should be pointed out, however that, if the swelling stresses during water adsorption were sufficiently great the cell wall would completely disperse and go into complete

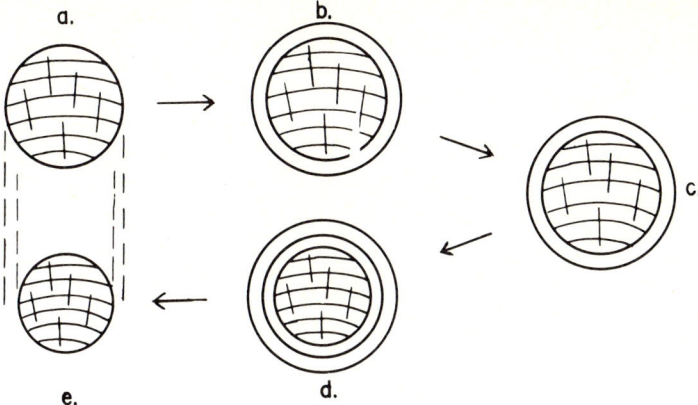

Fig. 3.30. Development of compression set in wood dowels which are restrained from swelling in the transverse directions by steel restraining rings. The residual compression set is shown as the difference in diameter of samples shown in *a.* and *e.*

solution in the water and the swelling would be infinite at unity relative vapor pressure, as it is for true solutions. This does not occur, fortunately, for two reasons. First, the water apparently does not penetrate to any extent into the crystalline portions of the cellulose, and second, the cellulose chains are not stressed to the point where they rupture during the normal course of swelling in water, thus restraining the basic integrity of the cell wall. It is also worth repeating that the *S*1 and *S*3 layers of the cell wall, with their microfibrils nearly perpendicular to the cell axis and to the *S*2 layer, restrict the swelling of this layer and thus reduce the external dimensional changes in the cell and of the gross wood.

4. Thermodynamics of Moisture Sorption by Wood

The interaction of water and wood is always accompanied by a change of heat or energy in the system. Many aspects of the interaction can be treated by the classical methods of thermodynamics, and it is the purpose of this chapter to do so. It is recognized that moisture sorption is not a perfectly reversible process since hysteresis and time-dependent phenomena are generally involved. However, our purpose here can be served by use of the simplifying assumption involved in reversible thermodynamics, keeping in mind the limitations of such an approach. In this chapter we will consider topics such as the heats of wetting and sorption and the interrelationships among these properties, as well as specific heat, swelling pressure, and hygroelastic and thermoelastic effects.

Differential Heat of Sorption

Moisture in wood occurs in three forms analogous to the three forms in which ordinary water exists. These are water vapor in air spaces in the cell cavities, capillary water in the cell cavities, and hygroscopic or bound water in the cell walls. Water-vapor molecules are at the highest potential energy level, capillary-water (sometimes referred to as free-water) molecules are at a lower potential energy level, and hygroscopic or bound-water molecules in the cell wall are at a still lower energy level.

Liquid-water molecules in the cell cavities are at a somewhat lower potential energy level than those of ordinary liquid water. The difference is small, however, and will be neglected (see Table 1.6 in Chapter 1); it is due to the force of capillary-water attraction which is small compared with the attractive force holding water in the cell wall. The water-vapor molecules in the cell cavities are in the same high energy state as water vapor outside the wood. The potential energy level of the hygroscopic or bound water in the cell wall is at the lowest level and represents the most stable condition. Bound water is similar in some respects to the frozen or solid state of ordinary water. However, it is different in that the sorbed water molecules are held with varying energies depending on the wood moisture content, whereas all of the molecules, in ice, are in nearly the same potential energy state.

127

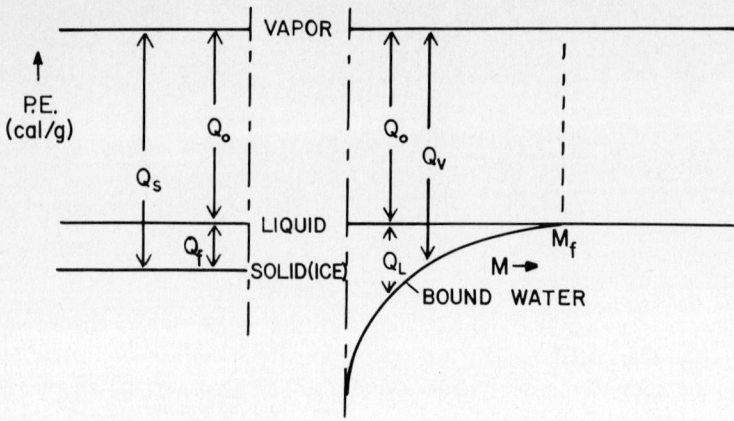

Fig. 4.1. Schematic diagram showing the relative potential energy levels of water vapor, liquid water, ice, and bound water in wood at different moisture contents M in terms of the energy Q (cal/g-water).

Figure 4.1 shows schematically how the potential energy levels of water in the various states compare. It is apparent from the figure that the potential energy level at the vapor state or ordinary water vapor is the same as that of water vapor in the wood and that this is the highest energy level. It is also clear that the potential energies of the liquid states are essentially the same inside as well as outside the wood; that is, capillary water in the cell cavity has nearly the same energy level as ordinary free liquid water. For this reason it is referred to as free water to distinguish it from bound or hygroscopic water in the cell walls. Furthermore, the lowest potential energy state for ordinary water is in the ice state while for water in wood the lowest energy state is in the hygroscopic or bound state which is the state of water in the cell wall. It is also clear from Figure 4.1 that the energy level for bound water is lowest near zero percent moisture content and highest at the fiber-saturation point M_f where it is essentially equal to that of liquid water.

The difference, $Q_v - Q_0$, where Q_v is the energy required to evaporate one gram of water from the cell wall and Q_0 is the energy required to evaporate one gram of water from the liquid state, is designated as Q_L, the differential heat of sorption of liquid water by wood. Q_L therefore is the additional heat energy, over and above the heat of vaporization Q_0 of free water, which must be supplied to evaporate one gram of water. It is analogous to the heat of fusion Q_f required to melt ice. At 50°C the value of Q_0 is 569 calories per gram of water, and the value of Q_L varies from 260 calories per gram of water for ovendry wood to zero at the fiber-saturation point. Thus, since $Q_v = Q_0 + Q_L$, it varies from $569 + 260 = 829$ at zero moisture content to $569 + 0 = 569$ calories per gram of water at M_f. Figure 4.2 shows the relationships of Q_L, Q_0, and Q_v with wood moisture content.

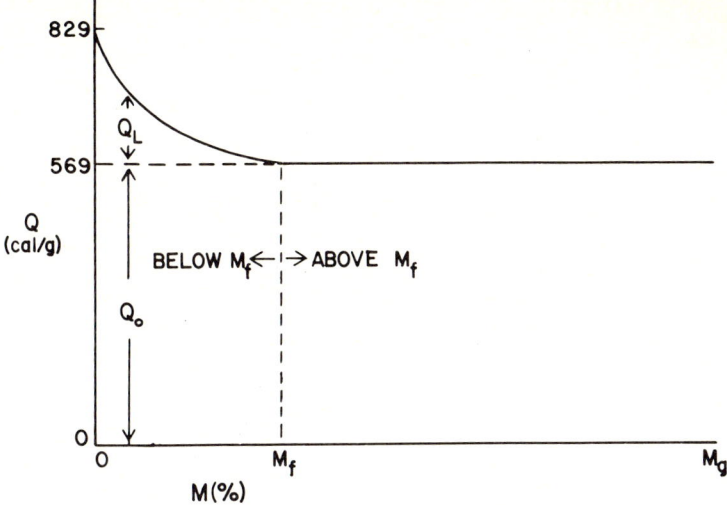

Fig. 4.2. Sorption energy Q (cal/g-water) of water in wood as a function of wood moisture content M below and above fiber saturation M_f.

There are two basic methods for finding Q_v or Q_L, just as there are two methods for finding the heat of fusion Q_f or the heat of sublimation Q_s of ice or the heat of vaporization Q_0 of water. One of these is the thermodynamic method using equation (1.16) of Chapter 1 or some variation of this equation. The second method is by direct calorimetric measurement. In this case the quantity of heat required to cause a change of state is measured by means of a calorimeter. For example, the heat of fusion or of melting of ice can be measured directly by measuring the quantity of heat required to melt a given amount of ice in an ice-water mixture contained in a calorimeter.

The procedure for finding Q_v by the thermodynamic method requires information on the variation with temperature of the vapor pressure p of wood at constant moisture content. When the logarithm of p is plotted against $1/T$, the reciprocal of absolute temperature, the value of Q_v is obtained by substitution of the slope, $d(\log p)/d(1/T)$, into the Clausius-Clapeyron equation (Chapter 1, equation 1.16) in the form

$$Q_v = -0.254(d(\log p)/d(1/T)) \text{ cal/g water.} \qquad (4.1)$$

In equation (4.1) the vapor pressure p of the wood must be measured at constant moisture content.

For example, Figure 4.3 shows the kinds of curves of $\log p$ against $1/T$ which might be expected at constant wood moisture contents of 5, 10, and 20 percent and at moisture contents above the fiber-saturation point, or say for green wood. These are not taken from real data and are therefore hypothetical. However, the

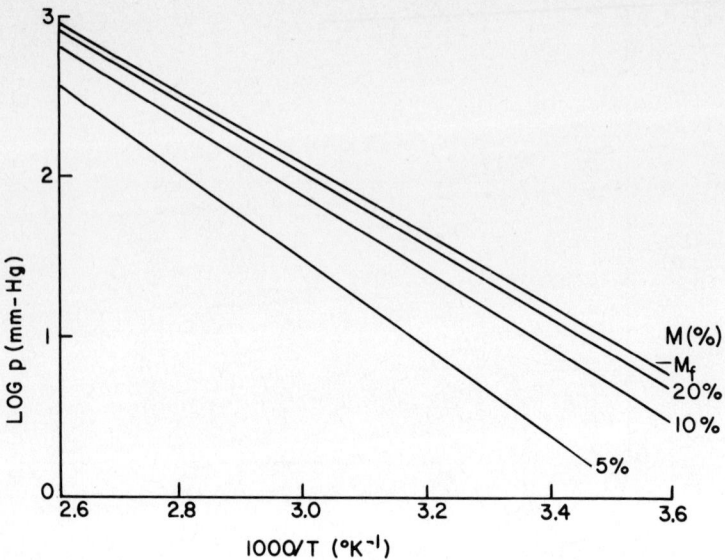

Fig. 4.3. Curves of the logarithm of vapor pressure p (mm-Hg) as a function of the reciprocal of absolute temperature T (°K) for wood of different constant moisture contents M (%).

general increase in the slopes of $\log p$ against $1/T$ is as anticipated as moisture content decreases, since Q_v increases with decreasing wood moisture content.

In order to obtain data from which to construct curves such as are shown in Figure 4.3 it is necessary to have sorption isotherms over a wide range of temperatures, such as those shown in Figure 4.4, which show curves only for 20°C and 80°C. Curves for intermediate temperatures should fall between these two curves and would be required in order to construct a diagram such as is shown in Figure 4.3. However, it is possible to calculate Q_v even when curves for only two temperatures are available, as in Figure 4.4, provided we know that the general relationships follow those given in equation (4.1); that is, that $\log p$ is linearly related to $1/T$.

Consider the sorption isotherms shown in Figure 4.4. The relative vapor pressure h at any given moisture content is higher at 80°C than at 20°C. The vapor pressure p at any temperature is equal to the relative vapor pressure (p/p_0) times the saturated pressure p_0 at the same temperature, or $p = p_0 (p/p_0)$. For example, from Figure 4.4, $p/p_0 = 0.54$ at 20°C and 0.74 at 80°C. The corresponding saturated vapor pressures p_0 are 17.54 and 355.1 mm Hg at 20°C and 80°C (see Table 1.1 in Chapter 1). Therefore, the vapor pressure p of the wood is $(17.54)(0.54) = 9.48$ mm Hg at 20°C and $(355.1)(0.74) = 263$ mm Hg at 80°C. The values of $\log p$ are 0.9768 at 20°C, and 2.4200 at 80°C. The reciprocal temperatures $1/T$ are 0.00341 and 0.00283, respectively. Using

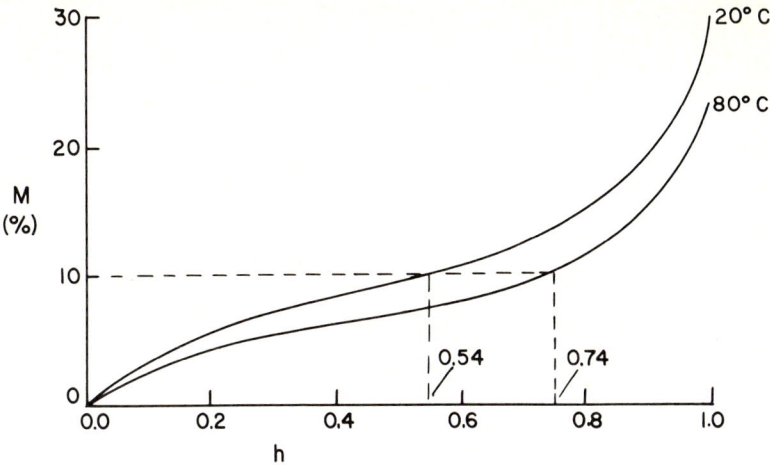

Fig. 4.4. Sorption isotherms for wood at 20°C and 80°C showing the increase in relative vapor pressure h with increasing temperature at constant wood moisture content M (%).

equation (4.1), Q_v can be evaluated as

$$Q_v = -0.254 \frac{(\log p_1 - \log p_2)}{(1/T_1) - (1/T_2)}$$

$$= -0.254 \, [(2.4200 - 0.9768)/(0.00283 - 0.00341)]$$

$$= 632 \text{ cal/g water}$$

The value of Q_L can be calculated at $Q_v - Q_0$. Therefore, since $Q_0 = 569$ cal/g water at the mean temperature of 50°C, and $Q_v = 632$ cal/g water, $Q_L = Q_v - Q_0 = 632 - 569 = 63$ calories per gram of water at 10 percent moisture content. Similar calculations can be made at other moisture contents from which a curve similar to that shown in Figure 4.5 can be obtained.

Q_L can also be calculated directly from the differences in the relative vapor pressures h or p/p_0 at different temperatures and constant wood moisture contents. This is accomplished by combining equation (4.1) with equation (1.16). These two equations are

$$Q_v = -0.254 \frac{d \log p}{d(1/T)}$$

$$Q_0 = -0.254 \frac{d \log p_0}{d(1/T)}.$$

Therefore, since $Q_L = Q_v - Q_0$

$$Q_L = -0.254 \frac{d \log (p/p_0)}{d(1/T)} \qquad (4.2)$$

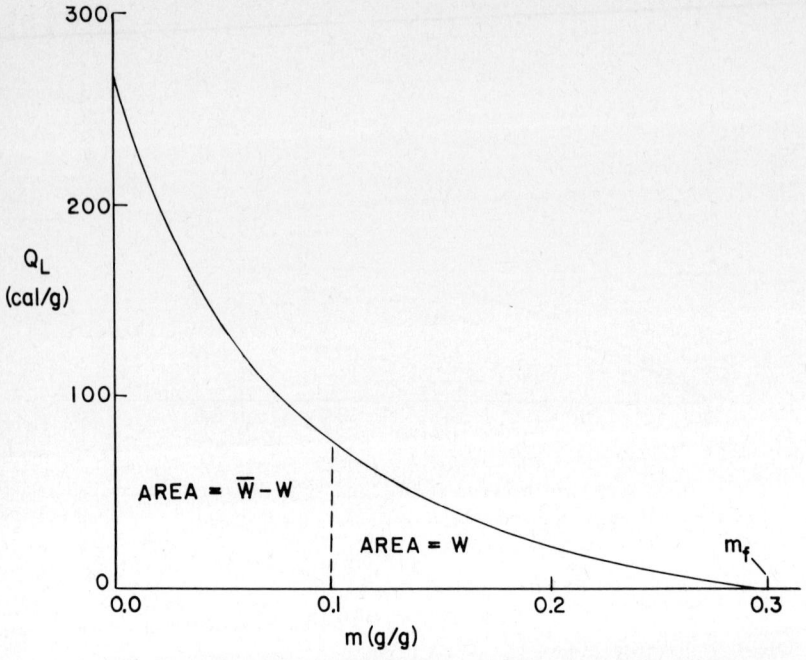

Fig. 4.5. Curve of differential heat of sorption Q_L (cal/g-water) against wood moisture content m (g/g). The area under the curve from $m = 0$ to 0.1 is equal to $\overline{W} - W$ (cal/g-wood), and the area from 0.1 to m_f is essentially equal to W at $m_i = 0.1$.

$$Q_L = +0.254 \frac{d \log (1/h)}{d(1/T)} \qquad (4.3)$$

When data for only two temperatures are available it may be more convenient to use equation (4.3) in the integral form. This can be written

$$Q_L = 0.254 \frac{T_1 T_2 \log (h_2/h_1)}{T_2 - T_1} \qquad (4.4)$$

where the subscripts refer to h corresponding to temperatures T_1 and T_2.

Stamm and Loughborough (1935) were the first to apply the thermodynamic method to the calculation of the differential heat of sorption Q_L for wood. Working with the sorption isotherms for Sitka spruce wood, they found a curve similar to that shown in Figure 4.5.

Weichert (1963) measured Q_L at a mean temperature of 62.5°C for spruce, beech, and "compressed" beech wood, using a vacuum sorption apparatus of high refinement. The results are shown in Table 4.1. The values for spruce agree closely with those found by Stamm and Loughborough (1935).

It is virtually impossible to measure the differential heat of sorption Q_L

Table 4.1. Differential Heat of Sorption Q_L
According to Weichert (1963)

MC (%)	Q_L (cal/g water)		
	Spruce	Beech	Compressed Wood
1	226.8	192.1	218.4
2	211.9	151.5	200.2
4	165.5	111.0	149.1
6	114.4	72.9	89.3
8	81.0	65.6	59.6
10	62.5	52.6	41.9
15	32.1	26.5	17.1
20	15.2	13.8	6.5

directly by calorimetric means. To do so would require the addition of a small amount of liquid water uniformly throughout a large enough sample of wood such that its moisture content would remain essentially unchanged. This is impossible to do experimentally with any assurance of accuracy. Theoretically, if a gram of water could be added to a very large mass of wood, the heat generated would be equal to Q_L at the moisture content of the wood.

However, it is possible to obtain Q_L indirectly from calorimetric experiments in which the energy released when wood thoroughly wetted with an excess of water is measured. The quantity which is measured directly in this type of experiment is called the integral heat of wetting, or simply the heat of wetting, designated by the symbol W. Its meaning and relationship to Q_L will be discussed next.

Integral Heat of Wetting

The heat of wetting W is defined as the heat generated per gram of dry wood when wood at some uniform initial moisture content is soaked in an excess of water such that it comes to a moisture content equal to or greater than the fiber-saturation point. It is somewhat analogous to the heat of solution generated when a hygroscopic substance is dissolved in water.

The heat of wetting for a particular wood depends on the initial moisture content of the wood before it is soaked in the water. For example, dry wood will produce more heat and therefore give a higher heat of wetting than wood which contains some moisture prior to soaking.

The method for measuring the heat of wetting W calorimetrically requires that the wood first be ground into small particles in order to minimize the time required to thoroughly saturate all of the wood with water. Before immersion in water the wood, after grinding, is conditioned to some desired, uniform moisture content, called the initial moisture content m_i ($=M_i/100$). The wood is then quickly immersed in water previously introduced into a calorimeter and the mixture or "slurry" of wood particles and water is stirred so as to obtain uniform moisture and temperature distribution. The heat generated by the reaction is calculated from the temperature rise of the calorimeter and the heat capacities of the various components in the calorimeter. This total heat, divided by the dry weight of the wood particles, is the heat of wetting W for the wood being measured. Many refinements are possible in the calorimetric heat of wetting measuring technique, designed to minimize heat-loss errors and to increase the precision of measurement.

The earliest measurements recorded for heat-of-wetting measurements on wood are those of Volbehr in 1896, according to Stamm (1964). Volbehr obtained the results shown in Table 4.2. It is clear from the table that the heat of wetting de-

Table 4.2. Heat of Wetting Data by Volbehr on Wood (According to Andersson 1952)

Initial Wood MC, M_i (%)	0	5	10	15	20	25	30
W (cal/g dry wood)	16.9	9.1	5.0	2.7	1.4	0.6	0.1

creases rapidly with an increase in wood moisture content. A convenient equation for relating W and m_i, of the form used by Cooper and Ashpole (1959), can be written as

$$\log W = A - Bm_i. \tag{4.5}$$

This equation gives a reasonably good fit to heat-of-wetting curves except at moisture contents approaching the fiber-saturation point. An approximate calculation over this range gives

$$\log W = 1.2335 - 5.408\, m_i \tag{4.6}$$

from the data of Table 4.2. This equation is convenient because only two constants are required, but it predicts that the heat of wetting becomes asymptotic to the zero value as m_i increases to large values. Strictly speaking, this may be true because of the finite but small heat generated by capillary sorption in the cell cavities. However, the use of this equation should be restricted to the hygroscopic range of moisture, below m_f, since the experimental data is confined to this region.

Integral Heat of Sorption

The integral heat of sorption is defined as the difference $\overline{W} - W$, where $\overline{W}$ is the total integral heat of wetting from the ovendry condition to saturation and

W is the heat of wetting from some initial moisture content m_i other than zero. For example, using the data of Volbehr from Table 4.2, the integral heats of sorption $\overline{W} - W$ are $16.9 - 16.9 = 0$ at 0 percent and $16.9 - 5.0 = 11.9$ (cal/g dry wood) at 10 percent.

The integral heat of sorption gives the energy generated per gram of dry wood when sufficient water is added to the dry wood to bring its moisture content up to the moisture content m_i or M_i. We shall discuss its significance in more detail later in this chapter. The integral heat of sorption is usually calculated from heat-of-wetting measurements of $\overline{W}$ and W, since this is more convenient experimentally than are direct measurements, which are difficult to make for reasons similar to those which make direct calorimetric measurements of Q_L almost impossible to carry out.

Relationship of Integral and Differential Heats of Sorption

The total heat of wetting $\overline{W}$ is equal to the total area under the curve in Figure 4.5. It can be defined mathematically as

$$\overline{W} = \int_0^{m_f} Q_L \, dm = 0.01 \int_0^{M_f} Q_L \, dM. \qquad (4.7)$$

The heat of wetting W is always less than $\overline{W}$ since it is defined as

$$W = \int_{m_i}^{m_f} Q_L \, dm \qquad (4.8)$$

where m_i is the initial wood moisture content. In Figure 4.5, W is the area under the curve to the right of the initial content corresponding to m_i. For example, if $m_i = 0.10$, then it is the area between $m = 0.10$ and 0.30, the fiber-saturation point.

The integral heat of sorption $\overline{W} - W$ therefore is the area to the left of m_i, or

$$\overline{W} - W = \int_0^{m_i} Q_L \, dm. \qquad (4.9)$$

The reverse operation can be performed to find Q_L at any wood moisture content if W or $\overline{W} - W$ are known as functions of m_i. Thus Q_L at the moisture content m_i is equal to the slope of the curve of the integral heat of sorption $\overline{W} - W$ against moisture content m. This slope can be obtained graphically from a plot of the data, or analytically as

$$Q_L = d(\overline{W} - W)/dm = -dW/dm \qquad (4.10)$$

since $\overline{W}$ is constant with respect to m.

If W is of the form given by equation (4.5), then it can be shown by substitution into equation (4.10) that

$$Q_L = 2.303 \, BW \qquad\qquad (4.11)$$

or

$$\log Q_L = C - Bm \qquad\qquad (4.12)$$

where $C = A + \log(2.303 \, B)$.

From equation (4.12) it appears that $\log Q_L$ is linearly related to moisture content m provided the linearity of $\log W$ and m is established. Figure 4.6 shows

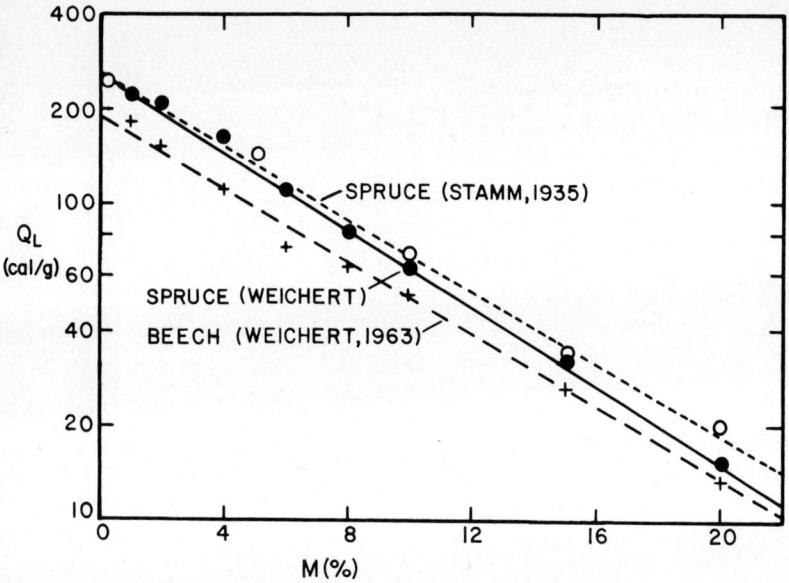

Fig. 4.6. Curves of the logarithm of Q_L (cal/g-water) against wood moisture content M for European spruce and European beech from data of Weichert (1963) and for Sitka spruce from data of Stamm and Loughborough (1935).

a plot of $\log Q_L$ against M using the data for two woods given in Table 4.1, as well as those of Stamm and Loughborough (1935).

The thermal quantities Q_L and W are related to each other as we have shown. However, they have different meanings in terms of what they signify about how water is held in wood. According to Rees, in Hearle and Peters (1960), the differential heat of sorption Q_L is a measure of the heat energy which must be supplied to sorbed water molecules to break the wood-water bond. It is a measure of the additional bond strength of the sorbed water for the OH (hydroxyl) groups, over and above the bonding strength of the water molecules for each other, which is measured by Q_0.

The total heat of wetting $\overline{W}$, on the other hand, is an indication of the total number of OH groups or sorption sites which are accessible to water. Wood contains regions which are accessible to water and also other regions such as the interior of crystallites which are inaccessible to water. There is evidence to indicate that the total hygroscopicity of a given wood, that is, its fiber-saturation point, is related to the total heat of wetting $\overline{W}$. For example, Raczkowski (1963) found that $\overline{W}$ was linearly related to the fiber-saturation point M_f such that the ratio $\overline{W}/M_f$ was essentially constant and equal to 0.8. In addition he found that $\overline{W}$ for sapwood was up to 30 percent higher than for heartwood and that alcohol-benzene extraction increased $\overline{W}$ by as much as 38 percent. This is probably related to the increase in the hygroscopicity of wood after such extraction (Higgins 1957) (see also Chapters 2 and 3).

Delgado (1970) found, for a number of Venezuelan woods segregated into six groups based on their hygroscopicity at 67 percent humidity, that the more hygroscopic woods showed higher total heats of wetting $\overline{W}$. A similar relationship has been found (Morton and Hearle 1962) for eight different textile materials of different hygroscopicities, expressed in the form $\overline{W} - W_{65} = 1.25\,M_{65}$, where $\overline{W} - W_{65}$ and M_{65} are the integral heat of sorption and moisture content of the material at equilibrium with 65 percent humidity.

The total integral heats of wetting $\overline{W}$ range between 15 to 20 calories or higher per gram of dry wood. Stamm (1964) has tabulated the values of $\overline{W}$ for a number of kinds of wood and also for wood constituents and other cellulosic materials. Some of these values are tabulated here for convenience in Table 4.3. Where more than one value is given in the table it indicates that smaller particle sizes yield higher values for $\overline{W}$, possibly because the structure is more accessible.

It is sometimes convenient, in calculating the efficiency of the drying process, to know the mean or average value of Q_L over a particular drying range. This

Table 4.3. Total Heat of Wetting ($\overline{W}$)
For Various Materials
(adapted from Stamm 1964)

Material	$\overline{W}$(cal/g dry wood)
Pine	17.9
Wood Flour (Sapwood)	19.3–19.8
Wood Flour (Heartwood)	18.1–18.4
Longleaf Pine (Sapwood)	17.6
Longleaf Pine (Heartwood)	16.7
Sugar Maple (Sapwood)	19.0
Sugar Maple (Heartwood)	19.6
Beech	16.6
W. Spruce (Alc.-Benz. Ext.)	18.2
Klinki Pine	18.9–20.5
Eucalyptus regnans	18.6
Cotton	9.7–12.3

value, which we will designate by $\overline{Q}_L$, can be calculated from the curve of Q_L against moisture content m as follows:

$$\overline{Q}_L = \frac{1}{m_f - m} \int_m^{m_f} Q_L \, dm \qquad (4.13)$$

or from the heat of wetting curve

$$\overline{Q}_L = W/(m_f - m) \qquad (4.14)$$

because of equation (4.12) which relates Q_L and W. In the case of drying, Q_L is the additional heat which must be supplied per gram of water evaporated from a given quantity of green wood over and above the normal heat of vaporization Q_0 of free water.

Free-Energy and Entropy Changes During Moisture Sorption

The differential heat of sorption Q_L is the total heat generated per gram of water absorbed by wood from the liquid state at a given wood moisture content. This heat is the total energy change involved during the sorption process and is believed to consist of two parts—one the change in the *work content or free energy* of the water sorbed and the other the change in *entropy or unavailable energy* of the water sorbed.

The amount of free energy change ΔG associated with the sorption of water by wood can be calculated at any given wood moisture content from the sorption isotherm, using equation (1.28) from Chapter 1, which states

$$\Delta G = (RT/18) \ln (p_0/p) \qquad (4.15)$$

in which ΔG is the free energy change or change in work content when one gram of water is taken up by wood at the moisture content at equilibrium with the relative vapor pressure p/p_0 at temperature T (°K). The change in free energy ΔG is undoubtedly related to the work involved in making sorption sites available by swelling the wood structure. It is reversible in the sense that the work is recoverable; that is, if the vapor is compressed back to the vapor pressure p_0 from the reduced vapor pressure p, the energy can be restored to the vapor.

The values of ΔG associated with the sorption process are plotted in Figure 4.7 together with the excess energy $T \Delta S$, defined as follows:

$$T \Delta S = Q_L - \Delta G \qquad (4.16)$$

where $T \Delta S$ is the change in entropy, a decrease in the case of adsorption of moisture.

The decrease of entropy associated with the sorption of water by wood is

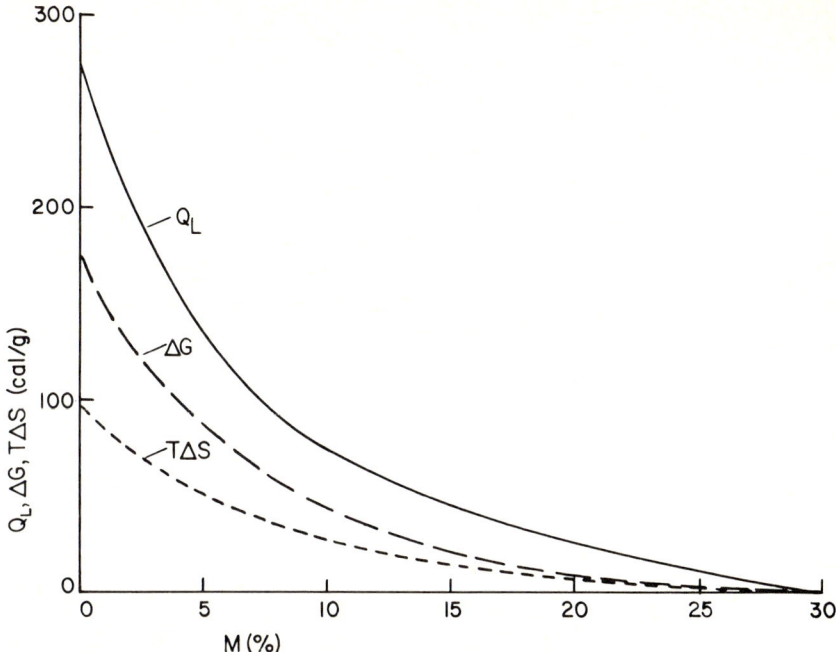

Fig. 4.7. Curves of differential heat of sorption Q_L (cal/g-water), free-energy change ΔG (cal/g-water), and of entropy change $T \Delta S$ (cal/g-water), when liquid water is taken up by wood at various moisture contents M (%).

attributed to the fact that the water in the cell wall is more ordered than is liquid water. It has been observed that the value of $T \Delta S$ of about 80–90 calories per gram of water absorbed is approximately equal to the entropy change when water freezes. When water freezes at $0°C$ all of the heat energy change Q_f is in the form of entropy change because of the increased order of ice molecules compared with those of liquid water.

As the relative vapor pressure h approaches zero, it should be noted that p_0/p, or $1/h$, approaches infinity and therefore that ΔG also becomes very large, as in equation (4.15). At extremely low humidities ΔG is larger than Q_L; they are equal at $h \approx 0.0001$. When ΔG is larger than Q_L it means that the entropy term $T \Delta S$ is negative; or, since ΔS is defined as the decrease in entropy when it is positive, a negative ΔS indicates an increase in entropy or an increased disorder in the sorbed molecules compared with liquid molecules. This is interpreted by Rees (in Hearle and Peters 1960) as being the result of the fact that the water molecules are so far separated from each other at very low wood moisture contents that they are in a greater state of disorder than those in the liquid state. It is likened to the expansion of a gas which results in an increase in entropy.

Specific Heat and Sorption

The specific heat c of any material may be defined as the calories of heat required to raise a one-gram mass by one degree Celsius. Ordinary liquid water is defined as having a specific heat of unity at $15°C$. The specific heat of dry wood, in common with that of most organic materials, increased markedly with temperature in the room-temperature region. This is attributed primarily to the fact (Skaar and Simpson 1968) that all three translational degrees of vibrational freedom of the individual atoms are not fully excited at these temperatures and are gradually brought into vibration as temperature increases. Most observers have found that the specific heat c_0 of dry wood increases linearly with temperature over limited temperature ranges as is shown in Table 4.4. The measure-

Table 4.4 Specific Heat c_0 (cal/g-deg) of Dry Wood
as Function of Temperature C ($°C$)

Wood	Source	Temperature Range	Equation for c_0
European Spruce	Volbehr (1896)	—	0.259 + 0.00121C
Several Species	Dunlap (1912)	0 to +112°C	0.266 + 0.00116C
European Spruce	Kühlmann (1962)	−60 to + 80°C	0.267 + 0.00118C
Spruce Pine	Koch (1969)	+60 to +140°C	0.265 + 0.00104C
Loblolly Pine	McMillen (1969)	+60 to +140°C	0.271 + 0.00095C
European Beech	Hearmon and Burcham (1956)	+30 to + 60°C	0.28 + 0.001C

ments by Volbehr (1896), Dunlap (1912), Kühlman (1962), and Hearmon and Burcham (1956) were made by standard calorimetric techniques, although the latter also used a quasi-steady-state thermal conductivity apparatus. Those by Koch (1969) and McMillen (1969) were made by means of a differential scanning calorimeter.

The specific heat c_m of moist wood is higher than that of dry wood because the specific heat of water is more than three times higher than that of dry wood. The simple method of mixtures has been used to calculate the specific heat of moist wood, based on the assumption that the wood and water contribute to the heat capacity of wet wood in proportion to their weighted mean heat capacities when calculated separately. In this case c_m can be calculated by

$$c_m = (c_0 + mc_w)/(1 + m) \qquad (4.17)$$

where m is the fractional moisture content based dry weight, and c_w, the specific heat of the sorbed water, is taken as unity.

Kelsey and Clarke (1956) have shown that the effective value of c_w in equation (4.17) is larger than unity if the integral heat of sorption $\overline{W} - W$ decreases with increasing temperature at any moisture content. This can be shown by recalling that $\overline{W} - W$ is really equal to the sum of the heat contents of the dry wood q_0 and of the water q_w, less the heat content q_m of the moist wood, all on a per gram of dry wood basis. In equation form this can be written

$$\overline{W} - W = (q_0 + q_w - q_m)/w_0 \tag{4.18}$$

where w_0 is the dry mass of the wood in grams and the units of q are in calories if $\overline{W} - W$ is in calories per gram. Differentiating both sides of equation (4.19) with respect to temperature T gives

$$d(\overline{W} - W)/dT = (1/w_0)(dq_0/dT) + (1/w_0)(dq_w/dT) - (1/w_0)(dq_m/dT). \tag{4.19}$$

Because of the following identities

$$c_0 = (1/w_0)(dq_0/dT) \tag{4.20a}$$

$$c_w = (1/w_w)(dq_w/dT) = (1/mw_0)(dq_w/dT) \tag{4.20b}$$

$$c_m = (1/w_m)(dq_m/dT) = \{1/[w_0(1 + m)]\}(dq_m/dT) \tag{4.20c}$$

equation (4.19) can be written as

$$d(\overline{W} - W)/dT = c_0 + mc_w - (1 + m)c_m \tag{4.21}$$

or

$$c_m = \frac{c_0 + mc_w - d(\overline{W} - W)/dT}{1 + m} \tag{4.22}$$

which differs from equation (4.17) in that it contains the extra term $d(\overline{W} - W)/dT$ in the numerator. If we designate Δc_w as the apparent increase in specific heat of the sorbed water, defined by

$$\Delta c_w = -(1/m)d(\overline{W} - W)/dT \tag{4.23}$$

then equation (4.22) can be written as

$$c_m = \frac{c_0 + m(c_w + \Delta c_w)}{1 + m}. \tag{4.24}$$

Kelsey and Clarke measured the heats of wetting W of klinki pine for different initial moisture contents from zero to 24 percent at three different temperatures and showed that the integral heat of sorption $\overline{W} - W$ decreased with increasing temperature above $m = 0.05$ (Figure 4.8). From their data they calculated $d(\overline{W} - W)/dT$ and Δc_w as functions of wood moisture content, obtaining the curves shown in Figure 4.9. The increase Δc_m in specific heat of the moist wood can be obtained from Δc_w as

$$\Delta c_m = \Delta c_w [m/(1 + m)] \tag{4.25}$$

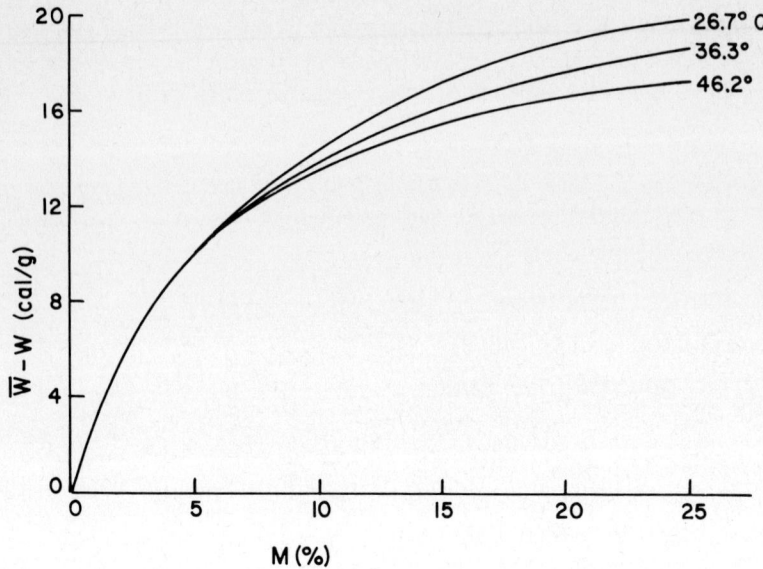

Fig. 4.8. Curves of integral heat of sorption, $\overline{W} - W$ (cal/g-wood), against wood moisture content $M(\%)$ at temperatures of 26.7°C, 36.3°C and 46.2°C for klinki pine (adapted from Kelsey and Clarke 1956).

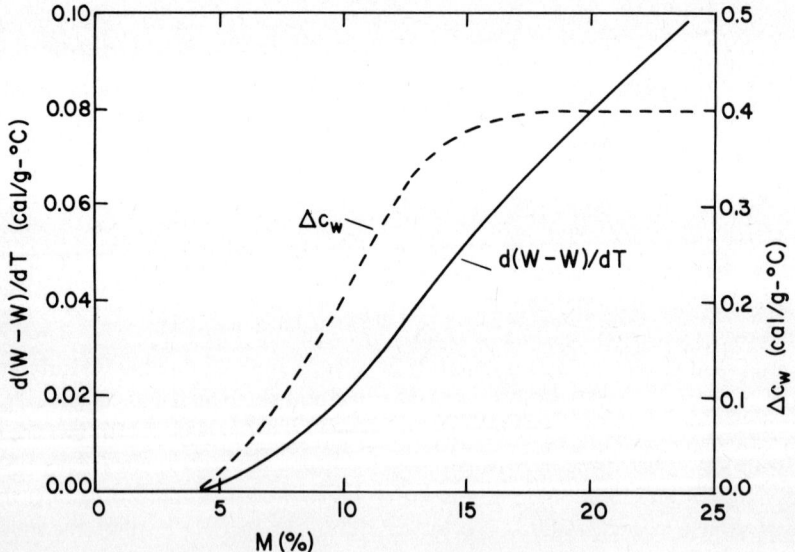

Fig. 4.9. Curves of $d(\overline{W} - W)/dT$ (cal/g-°C) and of Δc_m (cal/g-°C) as functions of wood moisture content $M(\%)$ for klinki pine (adapted from Kelsey and Clarke 1956).

Table 4.5. Increase Δc_m in Specific Heat of Moist Wood

Source	Condition of Wood		Excess Specific Heat Δc_m(cal/g°C)	
	m(g/g)	Temp(°C)	Meas. Directly	Calc. by Equ. (4.26)
Beech	0.107	30	0.02	0.02
(Hearmon and		40	0.02	0.02
Burcham 1956)		50	0.02	0.03
		60	0.04	0.05
	0.140	30	0.02	0.02
		40	0.03	0.03
		50	0.03	0.04
		60	0.06	0.05
	0.216	30	0.03	0.03
		40	0.04	0.04
		50	0.05	0.06
		60	0.06	0.07
	0.310	30	0.04	0.05
		40	0.05	0.06
		50	0.06	0.07
		60	0.09	0.09
Araucaria	0.05	27–46	—	0.00
klinkii	0.08	27–46	—	0.01
(Kelsey and	0.10	27–46	—	0.02
Clarke 1956)	0.12	27–46	—	0.03
	0.16	27–46	—	0.05
	0.20	27–46	—	0.07
	0.24	27–46	—	0.08

which, when combined with equation (4.23), gives

$$\Delta c_m = - [d(\overline{W} - W)/dT]/(1 + m). \qquad (4.26)$$

The values of Δc_m obtained by Kelsey and Clarke (1956) using equation (4.26) and Figure 4.9, are shown in Table 4.5.

Calorimetric measurements by Hearmon and Burcham (1956) of both the heat of wetting and specific heat over a moisture and temperature range confirmed the results of Kelsey and Clarke. Their results, also listed in Table 4.5, show the increase Δc_m in the specific heat of moist wood over the value calculated by the simple method of mixtures, given in equation (4.17) for a number of moisture contents and temperatures. They calculated c_m from direct calorimetric measurements and also from heat-of-wetting experiments using equation (4.22).

The curves of Stamm and Loughborough (1935) of Q_L against moisture content obtained by the isosteric or thermodynamic method also show a variation of Q_L with moisture content above 5 percent. Analysis of these curves, which can be converted into terms of $\overline{W} - W$ by use of equation (4.9), gives qualitative agreement with the findings of Kelsey and Clarke. However, the excess specific heat Δc_m is only about half the values found by Kelsey and Clarke. This may

be because of the difficulty in obtaining precise results using the thermodynamic
or isosteric method, as Kelsey and Clarke have pointed out.

Kühlman (1962) measured the specific heats of wood and of particleboard
using a quasi-steady-state thermal conductivity apparatus over the moisture
range from zero to 30 percent at temperatures between −60° and +80°C. His
results for moist wood are shown in Figure 4.10 and those for dry wood in
Table 4.4.

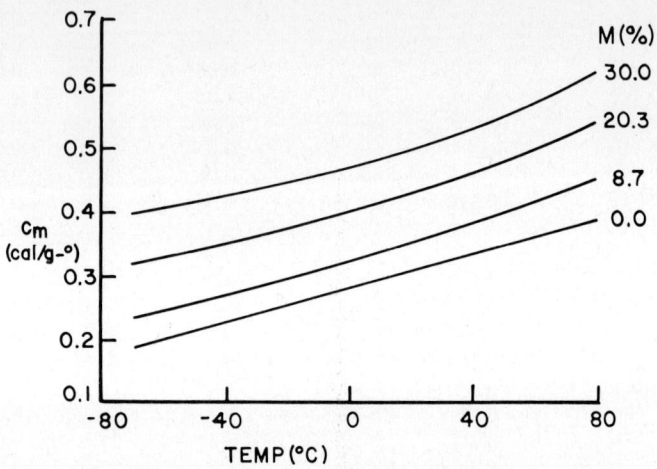

Fig. 4.10. Curves of specific heat c_m (cal/g-wood) against temperature (°C) for wood at
various moisture contents M (%) (adapted from Kühlman 1962).

Swelling Pressure of Wood

When wood takes up water, it swells to an extent which depends on the conditions under which the water is absorbed. One would expect, therefore, that
wood which is restrained from swelling should exert a stress in the direction
in which it is restrained from swelling. This is analogous to the swelling pressure
exerted by an osmotic solution against a restraining piston, such as shown in
Figure 4.11.

As described in Chapter 1, when an aqueous solution such as sugar in water is
separated from pure water by a semipermeable membrane, a driving force tends
to cause water to move into the solution from the pure water. This driving
force is known as the osmotic pressure and can be calculated from equation
(1.34) in Chapter 1.

In order to prevent migration of water through the membrane into the solution
a pressure must be applied to the solution which is equal to the osmotic pressure. This pressure can be applied by means of a piston, as shown in Figure 4.11,

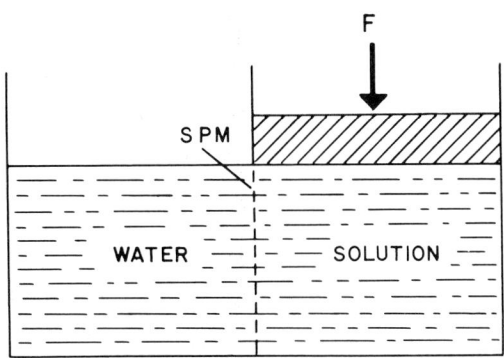

Fig. 4.11. Schematic diagram illustrating osmotic pressure showing the force F exerted by water against an aqueous solution from which it is separated by a semipermeable membrane (SPM).

with a force that will be sufficient to cause the pressure of the piston on the water to equal the osmotic pressure. This is one method of measuring osmotic pressure.

In the case of wood, an analogy exists to an aquaeous solution. The wood can be thought of as being partially dissolved in the water which is sorbed in the cell walls. As with a true solution this water is held at a vapor pressure p which is lower than the saturated vapor pressure p_0 of pure water. In fact wood is a material of the general class called gels which are similar to solutions in some respects but different in others.

Barkas (1949) has listed six characteristics of gels which are given here since they form a background for understanding the swelling pressure of wood as well as the relationship between stress applied to wood and its equilibrium moisture content:

1. *Gels are hygroscopic.* They hold moisture at a vapor pressure p at equilibrium with their surrounding atmosphere.

2. *Gels swell* when they take up water, to a greater or lesser extent than the volume of water taken up. Swelling may be anisotropic.

3. *Gels exert forces if restrained from swelling* when exposed to vapor pressures higher than their equilibrium vapor pressures. These forces may be different in different directions. Gels also sorb less moisture at a given vapor pressure when so restrained.

4. *Gels possess some rigidity* and can therefore withstand static shear stresses, unlike solutions which cannot withstand static shear stresses.

5. *Gels show limited sorption and swelling* in a saturated atmosphere. Thus they differ again from solutions which swell to infinite dilution in a water-vapor saturated atmosphere.

6. *Gels show sorption hysteresis* and thus differ from solutions which do not exhibit hysteresis.

Thus we see that gels are similar to solutions in their first three characteristics, except for the fact that they may be anisotropic; that is, they may have different properties in different directions, in which case they differ from solutions, which are isotropic. However, they differ from solutions in the last three characteristics since they have some rigidity, limited sorption, and swelling and show sorption hysteresis.

In order to measure the swelling pressure of a gel such as wood it is not necessary to provide a semipermeable membrane between the water and the gel. This is because the solute molecules, the wood molecules in this case, are essentially immobile and cannot migrate into the pure water outside of the cell wall (gel). However, the water molecules can migrate through the wood or gel structure and thereby pass into the gel. In order to measure the swelling pressure of the gel, it is only necessary to provide a restraint to swelling in the gel and expose the gel to liquid water.

In discussing the swelling pressure of wood we must distinguish between that of the cell wall or gel material and that of the gross wood, including the cell cavities or air spaces. The swelling pressure of the gross wood is limited by the wood's crushing strength which is a function of its density. Most of the experimental measurements of the swelling pressure of wood have been carried out on the gross wood, usually in one or more of the three principal structural directions. Perkitny (1963) has reviewed most of the important contributions on the swelling pressure of gross wood and has proposed standardized procedures for making and reporting such measurements. We will discuss first the swelling pressure of the gross wood, then the swelling pressure of the cell wall.

The swelling pressure of wood is usually measured by first drying the wood to a low moisture content followed by exposure to a humid atmosphere or to liquid water under conditions such that the wood is mechanically restrained from swelling in one or more directions. Usually the wood is restrained from swelling in one dimension only (Figure 4.12). Simpson and Skaar (1968), for example, reported curves of swelling pressure against time for red oak, of the form shown in Figure 4.13. These samples were initially dried to near one percent moisture content and then exposed to an atmosphere of 50 to 80 percent humidity at either of two temperatures, $80°F$ or $130°F$, under conditions where they were restrained from swelling either tangentially or radially. It is clear that the swelling pressures in each case reach a peak value and then decrease with time. This is believed to be caused by the simultaneous occurrence of swelling and relaxation of force with time. As the wood swells initially it exerts an increasing pressure against the restraining clamps. However, there is also a strong relaxation of stress because of the large rheological flow component in the wood. The initial rapid swelling causes the strong initial increase in swelling pressure, but the cu-

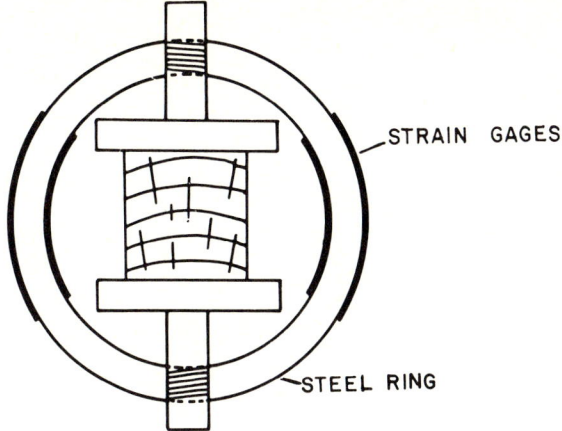

Fig. 4.12. Device for measuring unidirectional swelling pressure of wood. The strain gages measure the force exerted against the steel ring as the wood attempts to swell against the two restraining plates (adapted from Simpson 1966).

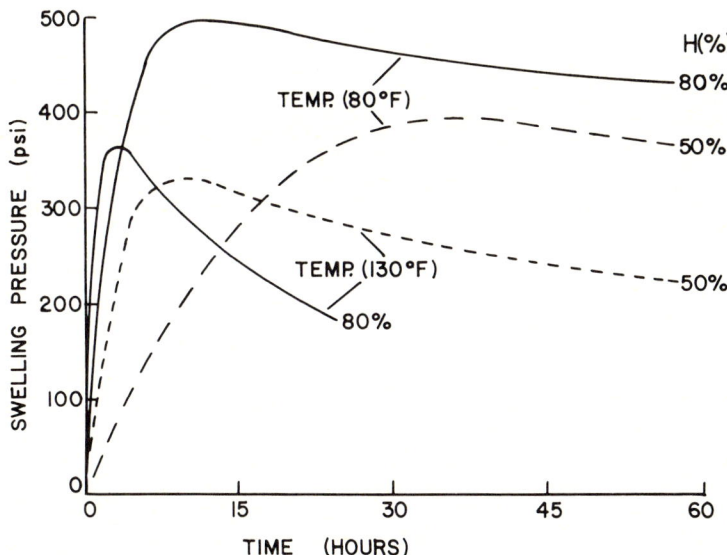

Fig. 4.13. Curves of tangential swelling pressure against time for dry red oak exposed to relative humidities H of 50 and of 80 percent at temperatures of 80° and 130°F (adapted from Simpson and Skaar 1968).

mulative effects of stress relaxation later dominate the process and the over-all swelling pressure decreases with time. Similar effects have been observed by Keylwerth (1962 and 1964) for wood restrained in any one of the three principal structural directions.

In the experiments described above, the swelling pressure was measured in a single direction in any given sample. In this case the maximum pressure which can be developed is limited by the compressive strength of the wood in the direction of restraint. Tarkow and Turner (1958) were able to measure the mean transverse swelling in two dimensions, using the arrangement shown in Figure 4.14. The restraining force was exerted by a steel ring into which the dry sample

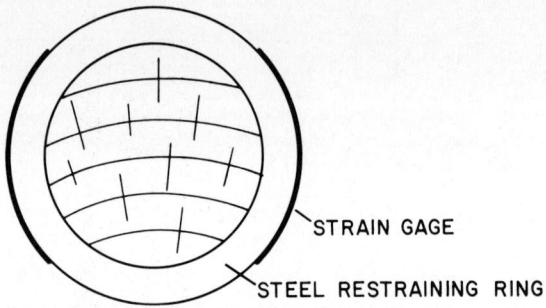

Fig. 4.14. Steel restraining ring with attached strain gages of the type used by Tarkow and Turner (1958) for measuring transverse swelling pressure of wood.

was force-fitted. On the outer perimeter of the ring were mounted two electrical resistance strain gauges which were used to measure the strain and hence the stress that the wood exerted in trying to swell within the ring. The dry wood was force-fitted into the steel ring and exposed to liquid water after first having been conditioned to equilibrium with 30 percent relative humidity. The swelling pressure or internal pressure on the ring was measured at different times after exposure to liquid water in order to trace the pressure variation with time.

According to the swelling-pressure equation (1.34) in Chapter 1,

$$\pi = (\rho RT / 18) \ln(p_0/p) \tag{4.27}$$

the maximum swelling pressure in the cell wall of wood should be about 1,560 atmospheres (see Table 1.5 in Chapter 1) at room temperature when wood at equilibrium with 30 percent humidity is exposed to saturated air. However, the values obtained experimentally were always considerably lower than this. This is attributed primarily to the fact that when wood is restrained from swelling externally, it swells into the cell cavities since these provide space for swelling. Tarkow and Turner could measure no appreciable swelling in the grain direction of the samples.

In order to minimize the effect of swelling into the cell cavities, Tarkow and Turner reduced the volume of cell cavities by densifying the wood to different degrees prior to measuring the swelling pressures. They found that the measured swelling increased rapidly with the extent of densification of the wood. The relationship appeared to be exponential as is shown in Figure 4.15, which shows

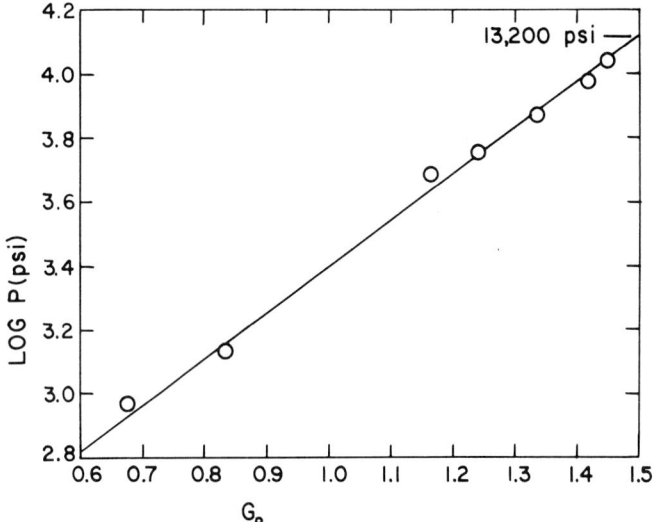

Fig. 4.15. Linear curve of the logarithm of swelling pressure (psi) as a function of dry wood specific gravity G_0, together with experimental points obtained from the measurements of Tarkow and Turner (1958).

the relationship between the logarithm of the maximum swelling pressure observed and the specific gravity of the samples, based on dry weight and volume. It is clear that the maximum apparent swelling pressure, as measured, increases exponentially with increasing specific gravity. The maximum value measured was 11,000 psi, about 750 atmospheres, obtained for a sample of 1.44 specific gravity. It therefore contained some void space into which swelling could take place. Extrapolation of the curve to specific gravity of 1.5, that of the dry cell wall, gives a swelling pressure of approximately 900 atmospheres (13,200 psi).

The discrepancy between theoretical (1,560 atm) and measured (900 atm) swelling pressures of the cell wall may be partially accounted for by the fact that the cell wall itself is somewhat compressible and some of the stress was used up in compressing the wood. Barkas (1949), in his discussion of stresses associated with moisture sorption, calls attention to this factor as one which must be used to modify equation (4.27). The swelling-pressure equation does not take into account the properties of the gel itself, but only those of the water. Therefore, quantitative work in which stress is related to changes in moisture content of a wood sample, the elastic properties of the wood must also be considered. This is the topic to which the Barkas Theory which relates moisture sorption in gels such as wood to the stress applied to the wood, is addressed, and will be treated next.

Mechanical Stress and Sorption (Hygroelastic Effect)

When wood is restrained from swelling and exerts a swelling pressure against the restraining structure it also comes to a lower EMC than wood which is unrestrained from swelling. This effect also operates in the reverse sense; that is, if a mechanical tension stress is applied to wood, its EMC is increased compared with that of an unstressed sample.

Barkas (1949) has treated this effect quantitatively, considering wood to be a hygroscopic gel and using the generalized Porter equation in the form

$$(\partial V/\partial m)_p dP_m = v\, dp \tag{4.28}$$

where V is the specific volume (cc/g) of the gel in terms of the swollen volume of the gel per unit dry mass (g), m (g/g) is the mass of water per unit mass of dry gel, dP_m is the increase in hydrostatic pressure required to raise the vapor pressure of the gel by the increment dp, and v (cc/g) is the specific volume of the water vapor. The term dV/dm actually represents the apparent specific volume (cc/g) of the sorbed water.

If the Porter equation is integrated, it reduces to the osmotic-pressure equation provided certain assumptions are made. Thus, letting ρ represent $\partial m/\partial V$, the apparent density (reciprocal of specific volume) of the sorbed water, and integrating between the limits p_1 to p_2, the Porter equation becomes

$$P_m = \int_{p_1}^{p_2} \rho v dp. \tag{4.29}$$

If the ideal gas law is now assumed to hold, and if the apparent density ρ of the sorbed water is considered to be constant, equation (4.29) becomes the familiar osmotic-pressure equation

$$P_m = (\rho RT/M) \ln (p_2/p_1). \tag{4.30}$$

Barkas cites four objections to integrating the Porter equation and applying the resulting simplified osmotic pressure equation to gels. First, the integrated equation neglects the rigidity of gels. Second, the integrated equation contains no provision for data concerning the gel itself but only of the water in the gel. Third, the integrated equation does not distinguish between the swelling pressure at constant moisture content from that at constant volume. And fourth, the vapor is presumed to obey the ideal gas law.

Because of these factors Barkas proposed a more generalized osmotic theory based on the Porter equation and which interrelated hydrostatic pressure P, specific volume V, moisture content m, and vapor pressure p. Figure 4.16 shows such a diagram calculated by Barkas from his generalized equation for the wood substance of Sitka spruce, using data of Stamm and Seborg (1935) on the specific volumes of this material over the full range of its sorption isotherm. This diagram shows that the moisture content m and the specific volume V of the cell wall are dependent not only on the vapor pressure p, or on h, but also on the external hydrostatic pressure P.

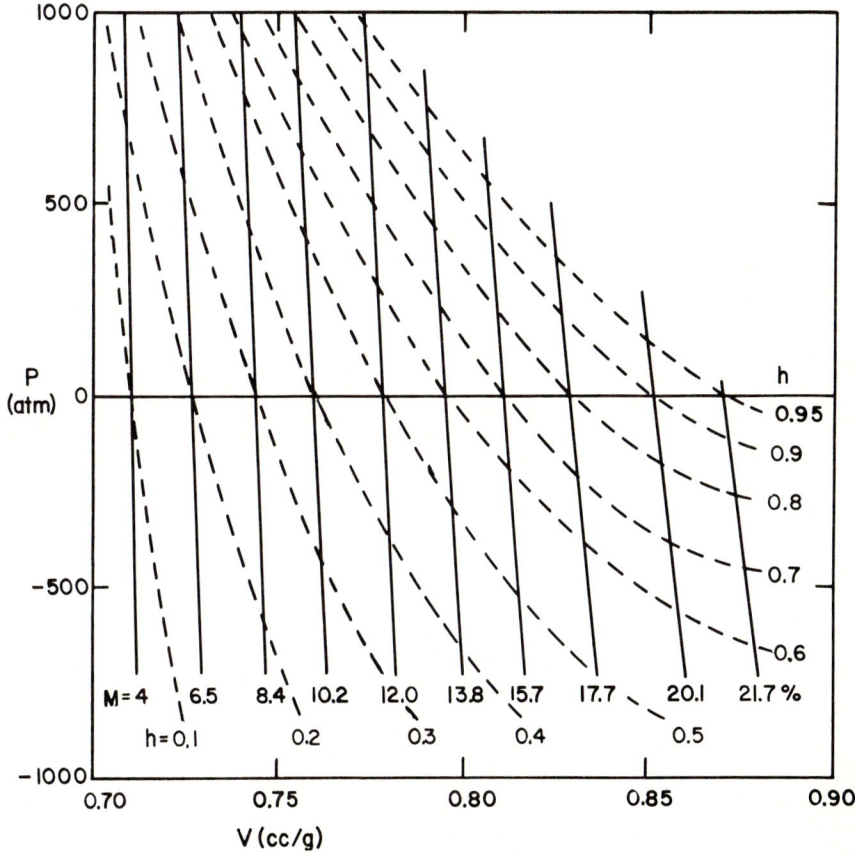

Fig. 4.16. Hydrostatic pressure P (atm)-specific volume V (cc/g) diagram for various constant moisture contents m and constant relative vapor pressures h as calculated for Sitka spruce (adapted from Barkas 1949).

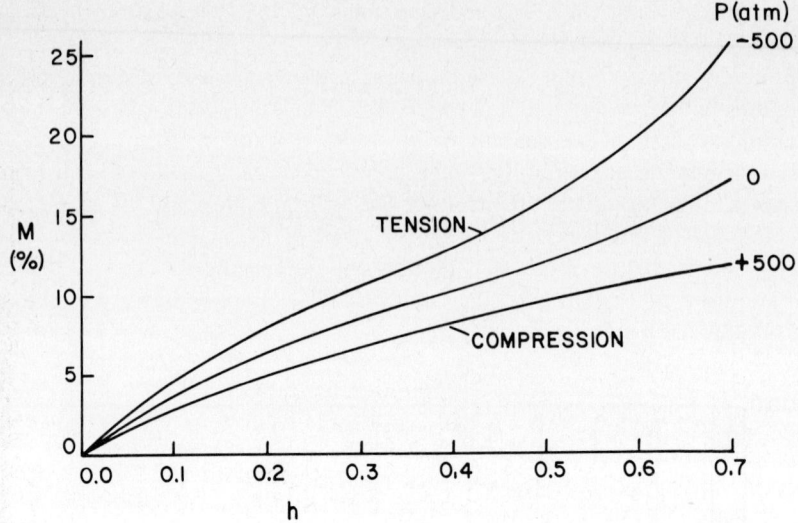

Fig. 4.17. Sorption isotherms for Sitka spruce as affected by hydrostatic stress P calculated from Figure 4.16.

The curves shown in Figure 4.16 can be replotted in the typical sorption isotherm form to show the effect of hydrostatic pressure or stress P on the sorption isotherms. For example, Figure 4.17 shows such curves—one for the stress-free condition, one for a positive hydrostatic pressure of 5×10^8 dynes/cm^2 (approximately 500 atm.), and one for a negative hydrostatic pressure (tension) of 5×10^8 dynes/cm^2. It is clear from these three curves that a hydrostatic pressure decreases the equilibrium moisture content at a given vapor pressure and that a hydrostatic tension increases the equilibrium moisture content.

Barkas has pointed out that even in the absence of external restraint there is an internal restraint to swelling because of the nature of the cell-wall structure. This internal restraint, caused by at least two factors, effectively reduces the hygroscopicity of the wood in the same way as does an external pressure or stress. One of these internal restraining factors is attributed to the differences in fibril orientation in the $S1$ and $S3$ layers of the cell wall compared with the $S2$ layer (Figure 3.6). These fibrils have their long axes nearly perpendicular to those of the bulky $S2$ layer and therefore reduce the external swelling and sorption. A second restraining factor is the presence of interfibrillar bonds which limit the swelling between fibrils. According to the fringe-micellar theory, for example, these bonds would consist of the cellulose chain molecules which interweave among the crystalline regions of the cell wall, thus restricting the degree of swelling in the intercrystalline or amorphous regions.

Figure 4.18, taken from Barkas (1949), shows a small portion of a P, V, m, p

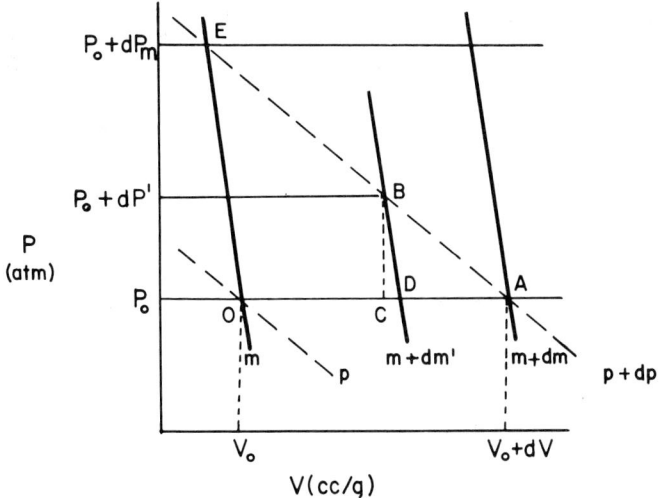

Fig. 4.18. Small portion of the P, V diagram of Figure 4.16 greatly enlarged (adapted from Barkas 1949).

diagram, such as shown in Figure 4.16, but for a gel which is assumed for convenience to be homogeneous, isotropic, and perfectly elastic. Before formally deriving the equations used by Barkas it is instructive first to consider the principles involved as illustrated in Figure 4.18. The gel is at a fractional moisture content m (g/g), at equilibrium with vapor pressure p, and having a specific volume V_0 (cc/g) based on swollen gel volume and dry gel mass, at the reference hydrostatic pressure P_0, which may be atmospheric pressure, for example. This condition is defined by point O in the diagram.

If the gel is now exposed to a higher vapor pressure, $p + dp$, and is allowed to swell freely at constant hydrostatic pressure P_0, it swells to a new specific volume $V_0 + dV$ as it adsorbs an additional moisture increment dm. The new equilibrium condition corresponds to point A on the diagram.

If an increment of hydrostatic pressure or stress dP' is now applied, maintaining a constant vapor pressure $p + dp$, the moisture content reduces to $m + dm'$, lower than $m + dm$, and the specific volume V decreases. This new stress-induced equilibrium corresponds to point B.

In order to reduce the moisture content to m at the vapor pressure $p + dp$ a total stress increment dP_m must be applied. At this equilibrium condition (point E) the specific volume is lower than at point O, even though m is the same. The difference is attributed to the compressibility, or to its reciprocal, the bulk modulus k_m of the gel at constant moisture content, which is directly proportional to the slope $(\partial P/\partial V)_m$ of the constant m line OE. It should be noted that this is much steeper than the slope $(\partial P/\partial V)_p$ of the line of constant p (essentially parallel to AE). Therefore the bulk modulus k_p at constant vapor pressure is much lower than k_m.

The derivation as given by Barkas begins by considering the equilibrium condition represented by point O in Figure 4.18. If the gel absorbs dm grams of water (per gram of dry gel) under constant hydrostatic stress P, its volume increases from O to A, and

$$OA = dV = (\partial V/\partial m)_p dm \qquad (4.31)$$

and its vapor pressure p increases to $p + dp$.

If a partial restraint is imposed on the gel, the hydrostatic stress P_0 increases to $P_0 + dP'$ along the path OB. At point B its vapor pressure is $p + dp$, as before, but it has taken up a reduced amount of moisture dm' and its volume increase dV' is now smaller than dV, its volume increase when unrestrained. Thus $OC = dV' = (\partial V/\partial m)' dm'$.

The volumetric compression due to the stress dP' is $OD - OC$ or CD, where OD is the volume increase the gel would have exhibited if it had taken up dm' grams of moisture at constant stress P_0.

The volumetric strain associated with the increment of moisture dm' under the stress increment dP' is CD/V, or

$$CD/V = [(\partial V/\partial m)' dm' - (\partial V/\partial m) dm'] /V. \qquad (4.32)$$

The volumetric or hydraulic stress dP', associated with restrained sorption, can also be calculated from the differences in the moisture sorption increments dm and dm' if it is assumed that the soprtion increment difference $(dm - dm')$ is directly proportional to the volume difference, $(dV - dV')$. Thus, in Figure 4.18 this means that the distances OD and OA are proportional to dm' and dm, or that

$$(OA - OD)/OA = DA/OA = (dm - dm')/dm. \qquad (4.33)$$

Furthermore, it is assumed that the triangles OEA and DBA are similar, which means that the volume strain for a given vapor-pressure increment dp is directly proportional to the hydrualic stress applied in restraining the swelling associated with the increment of vapor pressure. Thus, by the law of similar triangles

$$DA/OA = dP'/dP_m \qquad (4.34)$$

and from the equation (4.33)

$$dP'/dP_m = (dm - dm')/dm \qquad (4.35)$$

or

$$dP' = [1 - (dm'/dm)] dP_m. \qquad (4.36)$$

But, from the Porter equation, $dP_m = (v\, dp)/(\partial V/\partial m)_p$, and $dP' = v\, dp(1 -$

$(dm'/dm))/(\partial V/\partial m)_P$ or

$$dP' = \frac{v\,[(\partial p/\partial m)' - (\partial p/\partial m)]\,dm'}{(\partial V/\partial m)_P} \qquad (4.37)$$

which is the *stress* associated with restrained sorption. The bulk modulus k_m is equal to the hydraulic stress in equation (4.37) divided by the volume strain in equation (4.32). Thus, making this substitution and rearranging, k_m becomes

$$k_m = \frac{Vv\,[(\partial p/\partial m)' - (\partial p/\partial m)]}{(\partial V/\partial m)_P\,[(\partial V/\partial m)' - (\partial V/\partial m)]}\,. \qquad (4.38)$$

Equation (4.38) can be applied to isotropic gels or to solutions. It relates the bulk modulus k_m at constant moisture content to the difference in the slopes of the sorption isotherm for a gel in the restrained $(\partial p/\partial m)'$ and unrestrained $(\partial p/\partial m)$ condition and to the corresponding differences in the differential volume change in the restrained $(\partial V/\partial m)'$ and unrestrained $(\partial V/\partial m)$ conditions.

For an anisotropic gel such as wood there are shear strains involved even when it is subjected to hydrostatic pressures which are uniform on all surfaces. This is true both at the macroscopic or gross level, such as in a block of wood, and also at the microscopic or cell-wall level, as Barkas points out. Therefore, equation (4.38) cannot be applied rigorously. However, it is possible to apply it in a modified form for mechanical stresses resulting from unidirectional restraints applied normal to the principal orthotropic axes of wood such as in the radial or tangential directions during swelling. In this case, if Poisson effects and shear stresses are neglected, and if $(\partial V/\partial m)$ and $(\partial V/\partial m)'$ are taken to be equivalent to $(V/x)\,(\partial x/\partial m)$, and $(V/x)\,(\partial x/\partial m)'$, respectively, equation (4.38) can be modified to

$$E_m = \frac{v\left[\left(\dfrac{\partial p}{\partial m}\right)' - \dfrac{\partial p}{\partial m}\right]}{V\left(\dfrac{1}{x}\right)\left(\dfrac{\partial x}{\partial m}\right)_P\left[\left(\dfrac{1}{x}\right)\left(\dfrac{\partial x}{\partial m}\right)' - \left(\dfrac{1}{x}\right)\left(\dfrac{\partial x}{\partial m}\right)\right]} \qquad (4.39)$$

where E_m is Young's modulus in the direction x of swelling, $(1/x)(\partial x/\partial m)$ is the unrestrained rate of swelling in the x direction and $(1/x)\,(\partial x/\partial m)'$ is the restrained swelling in the x direction, the other terms having the same meaning as in equation (4.38).

In order to apply equations (4.38) or (4.39) to calculate the bulk modulus k_m or Young's modulus E_m, it is necessary to have the slope of the sorption isotherm in the unrestrained $(\partial p/\partial m)$ and in the restrained condition $(\partial p/\partial m)'$, and also the volumetric or linear swelling in the unrestrained $(\partial V/\partial m)$ or $(1/x)$ $(\partial x/\partial m)$ and restrained $(\partial V/\partial m)'$ or $(1/x)(\partial x/\partial m)'$ conditions. Barkas (1945) gives data obtained from measurements on spruce cubes which were restrained from swelling in the tangential directions during adsorption from equilibrium

Table 4.6. Data Obtained by Barkas (1945)
on Sitka Spruce for Calculating Young's Modulus
in Tangential Direction

mean value of h	0.398
Δh	0.279
$p_0 v$ (dyne-cm/g)	3.41×10^9
mean value of V (cc/g dry wood)	2.581
Δm unrestrained (g/g)	0.0396
$\Delta m'$ restrained (g/g)	0.0382
$(1/x)\,(\partial x/\partial m)$ unrestrained $(g/g)^{-1}$	0.386
$(1/x)\,(\partial x/\partial m)'$ restrained $(g/g)^{-1}$	0.120
Young's Modulus E_m (dynes/cm^2)	3.32×10^9

with h at 0.258 to equilibrium with h at 0.538. These data, together with those obtained on unrestrained samples exposed to the same conditions, are given in Table 4.6. It should be noted that the terms $(\partial p/\partial m)$ and $(\partial p/\partial m)'$ in equations (4.38) or (4.39) can be written in terms of relative vapor pressure h, in place of p, in which case the term $p_0 v$ is substituted for v.

Substitution of the data into equation (4.39) gives the following

$$E_m = \frac{(3.41 \times 10^9)[(0.279/0.0382) - (0.279/0.0396)]}{(2.581)(0.386)[0.120 - 0.386]}$$
$$= 3.32 \times 10^9 \, dynes/cm^2.$$

This value agreed reasonably well with measurements made in matched material conditioned to equilibrium with $h = 0.40$ using a mechanical testing machine.

In more recent work Bello (1968) has calculated Young's moduli in the radial and tangential directions from measurements on several hardwood species which were restrained from transverse swelling by use of steel rings into which samples were inserted (Figure 4.14) prior to exposure to a higher humidity. In this case the values of E obtained were in the right order or magnitude. He also found that the EMC was 1.44 percent lower on the average for restrained than for unrestrained samples when dry samples were exposed to 87 percent humidity, and 0.51 percent lower when exposed to 58 percent humidity, both at 77°F.

Equations (4.38) and (4.39) relate the elastic moduli of gels to the differences in swelling and water sorption between samples which are partially restrained from swelling compared with those which are unrestrained. It has been indicated during the development of these equations that the application of a compressive mechanical stress decreases sorption while a tension stress increases sorption. This is clear from Figures 4.16, 4.17, and 4.18 and also from the Porter equation itself.

It is more convenient experimentally to measure the effect of a constant mechanical stress, applied continually in one direction only, on the equilibrium moisture sorption at constant vapor pressure. For this case the Porter equation can be written in the form

$$(\partial V/\partial m)_p = v(\partial p/\partial P)_m \qquad (4.40)$$

or, more conveniently for a uniaxial stress σ applied in the x direction only, as

$$(V/x)(\partial x/\partial m)_\sigma = v(\partial p/\partial \sigma)_m . \qquad (4.41)$$

From the rules of partial differentiation it can be shown that

$$(\partial p/\partial \sigma)_m = -(\partial m/\partial \sigma)_p (\partial p/\partial m)_\sigma . \qquad (4.42)$$

Substituting this into the preceding equation and rearranging gives

$$(\partial m/\partial \sigma)_p = -(V/v) (1/x) (\partial x/\partial p)_\sigma . \qquad (4.43)$$

Equation (4.43) can also be written in the form

$$\left(\frac{\partial m}{\partial \sigma}\right)_p = - \left[\frac{(1+m)}{\rho \, v \, p_0}\right] \left[\left(\frac{1}{x}\right)\left(\frac{\partial x}{\partial h}\right)\right]_\sigma \qquad (4.44)$$

by recalling that $V = (1+m)/\rho$, where ρ is the density at moisture content m.

Equation 4.44 is in a convenient form for comparing experimental results with those of theory. It states that the change in moisture content m per unit of applied stress is directly proportional to the moisture content m and to the change

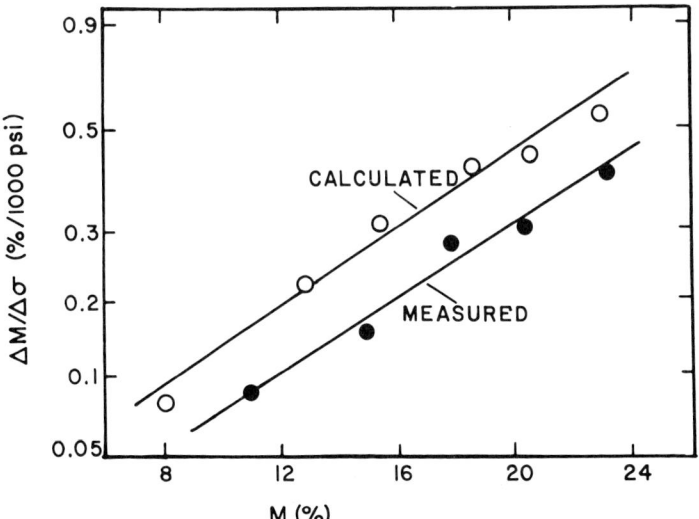

Fig. 4.19. Curves of $(\Delta M/\Delta \sigma)_p$ (%/1,000 psi) on a logarithmic scale against wood moisture content M calculated from equation (4.44) and also measured directly (adapted from Simpson 1971).

in swelling rate with vapor pressure and inversely proportional to the wood density ρ.

Simpson (1971) has carried out experiments in which he measured the effect $(\partial m/\partial\sigma)_p$ for compression and tension stresses on red oak in both the tangential and radial directions over a range of wood moisture contents. He also calculated the anticipated value from equation (4.44). A typical pair of curves showing both experimental and calculated values are shown in Figure 4.19. The experimental values were always smaller than the calculated values, but the variation with wood moisture content was similar.

Thermal Expansion and Contraction of Wood

It is generally observed that most materials expand when they are heated and contract when cooled, unless some change of state occurs. This behavior can be understood qualitatively in terms of Figure 4.20, which shows the typical shape of a potential energy curve between two adjoining atoms as a function of their spacing. The potential energy is a minimum at the equilibrium spacing x_0 which represents the energy when there is no kinetic energy of vibration. However, in

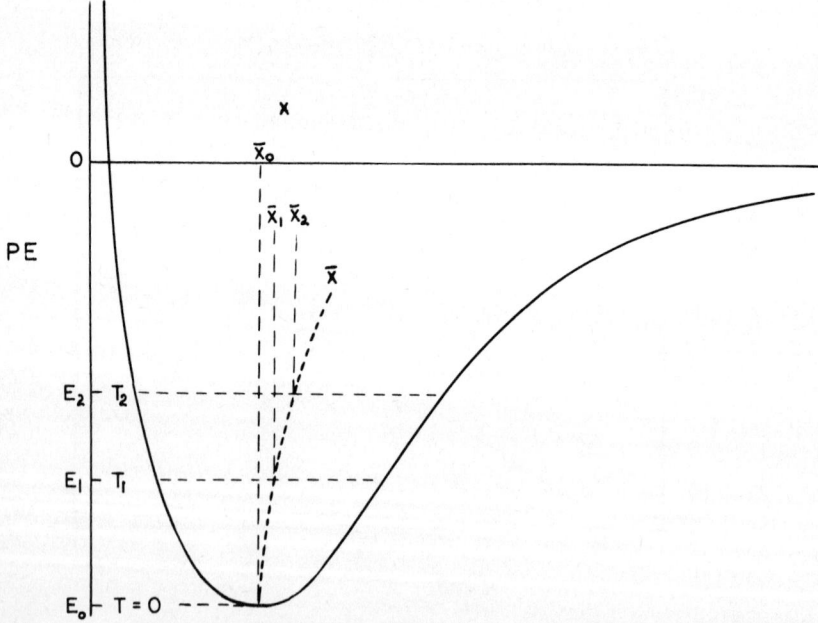

Fig. 4.20. Potential energy (PE) curves as a function of spacing x between two adjoining atoms.

general there is kinetic energy in the system, and the atoms vibrate with respect to each other with a kinetic energy proportional to the temperature. Thus at temperatures T_1 and T_2 their energies might be E_1 and E_2. Because of the asymmetry of the potential energy curve the mean spacings x_1 and x_2 are increasingly greater as temperature increases. Since the spacing between each pair of atoms increases with temperature, as shown schematically by the broken line in Figure 4.20, the linear dimensions of a solid composed of many atoms also increases with temperature. Furthermore the expansion with temperature may be nonlinear, increasing with temperature.

The coefficient of linear expansion α can be defined as

$$\alpha = \left(\frac{x_2 - x_1}{x_1} \right) \left(\frac{1}{\Delta T} \right) \tag{4.45}$$

or more generally, for a solid of linear dimension L

$$\alpha = (\Delta L / L) / \Delta T \tag{4.46}$$

where L is equivalent to the sum of all the interatomic spacings in one direction at the initial temperature, and ΔL is the total change in L over the temperature range ΔT. It should be kept in mind that the shape of the potential energy curve in Figure 4.20 is affected not only by the forces between the two atoms but also by their interaction with all neighboring atoms in a solid. Furthermore, the shape of the curve in an anisotropic material such as wood is expected to be different at different orientations with respect to the cell wall and to the grain of the gross wood. It is also anticipated that it would vary from one location to another within the cell wall because of local variations in chemical and physical structure.

Historically it is interesting to note the pioneering work on thermal expansion of wood carried out by James Joule (1859), who measured an α of 4×10^{-6} deg^{-1} for pine wood along the grain. Joule's objective in measuring α for wood was to relate this property to the thermoelastic effect which is discussed in the next section.

Weatherwax and Stamm (1946) have made a comprehensive study of the linear thermal-expansion coefficient α of dry wood. Their measured parallel-to-grain values of α_1 ranged from about 3×10^{-6} to 4.5×10^{-6} per °C, essentially independent of wood density. The transverse coefficients were much higher and were density dependent, averaging for the various species

$$\alpha_{0,r} \approx (50 \, G_0) \, 10^{-6} \tag{4.47}$$

$$\alpha_{0,t} \approx (70 \, G_0) \, 10^{-6} \tag{4.48}$$

where $\alpha_{0,r}$ and $\alpha_{0,t}$ are the coefficients in the radial and tangential directions and G_0 is the specific gravity in the dry state.

The reason for the low value for longitudinal expansion compared with the transverse expansion is probably related to the nature of the chemical bonds in the different directions. Those composing the cellulose chain molecule links in the longitudinal direction, which is the direction of their primary orientation, are strong covalent bonds characterized by deep and narrow potential energy curves. Those linking the adjacent chains together in the transverse direction are largely hydrogen bonds which are characterized by shallow and wide potential energy curves. One would therefore anticipate that the effect of temperature would be much larger in the transverse than in the longitudinal direction. It should be noted that there is an analogy between the directional behavior of thermal expansion and that of moisture expansion or swelling. As is the case with moisture expansion the longitudinal thermal expansion α_l is lowest, followed by radial α_r and tangential α_t expansion. Furthermore, in both thermal and moisture expansions the radial and tangential expansions are density dependent whereas the longitudinal expansions appear to be independent of wood density.

The thermal dimensional changes induced in wood containing moisture are more complex than for dry wood. Schirp and Kübler (1968) describe at least four different mechanisms which may operate to affect thermal dimensional changes in moist wood. Each mechanism is effective over a different range of wood moisture contents. In the air-dry range, well below the fiber-saturation point, the first mechanism—the normal thermal expansion or contraction previously described—occurs. However, its magnitude may be different than that of dry wood because of the water in the cell wall. The second mechanism, which occurs at moisture contents somewhat above fiber saturation, is related to the change in fiber-saturation point with temperature. A third mechanism, which occurs in wood in which the cell cavities are nearly full of water, is related to the expansion in cell-cavity water when it freezes. The fourth mechanism, which is effective only if the cell cavities are essentially full of water, is caused by the large thermal-expansion coefficient of water in the cell cavities. It is particularly significant in the region between zero and 4°C where the expansion coefficient for water is negative. In addition to the four reversible mechanisms described above there is also an irreversible dimensional change that takes place when green wood is heated in water. The four reversible mechanisms will be discussed first, followed by a description of the irreversible dimensional changes associated with heating green wood.

There is not much published information available on the normal thermal expansion of wood containing hygroscopic water. If certain assumptions are made with respect to the thermal-expansion behavior of hygroscopic water, however, it is possible to calculate hypothetical values of α for moist wood. For example, if it is assumed that the thermal expansion of bound water sorbed in the cell wall is equal to that of free liquid water, the method of mixtures can be applied as was done for calculating the specific heat of moist wood. It is recognized, as was true for the case of specific-heat calculations, that this assumption may not be valid. If it is also assumed that all of the expansion is in the transverse direction in the wood and that the expansion of the sorbed water is in the same ratio, 70 to 50 or 0.58 to 0.42, as that for dry wood in the tangential and radial directions, then, according to Skaar and Simpson (1968), the expected values of $\alpha_{t,m}$ and $\alpha_{r,m}$, the tangential and radial thermal-expansion coefficients for moist wood, are given by

$$\alpha_{t,m} = \frac{70\,G_0(10^{-6}) + 0.58\,G_0 m\alpha_{v,w}}{1 + 0.67\,G_0 m} \tag{4.49}$$

$$\alpha_{r,m} = \frac{50\,G_0(10^{-6}) + 0.42\,G_0 m\alpha_{v,w}}{1 + 0.33\,G_0 m} \tag{4.50}$$

where $\alpha_{v,w}$ is the coefficient of volumetric thermal expansion for water equal to approximately 200×10^{-6} per $°C$ at room temperature.

The second mechanism which causes dimensional changes in moist wood with temperature is associated with the change in the water-holding capacity or fiber-saturation point of the cell wall with temperature. This mechanism is pronounced at wood moisture contents at or above the fiber-saturation point. The fiber-saturation point decreases with increasing temperature above the freezing point of water, whereas the reverse is true at lower temperatures (Kübler 1962). It is therefore anticipated that wet wood should have a negative temperature coefficient above freezing because of the fact that water migrates out of the cell wall into the cell cavity with subsequent moisture shrinkage as temperature increases. This is caused essentially by the fact that the vapor pressure of wood at constant wood moisture content increases more rapidly with temperature than does the vapor pressure of free water or of the capillary water in the cell cavities.

If the moisture-shrinkage behavior and the rate of change of fiber-saturation point with temperature are known, it is possible to develop equations which can be used to estimate the magnitude of this effect in terms of the negative thermal expansion coefficient $\alpha_{t,f}$ and $\alpha_{r,f}$ where the subscripts t and r refer to the tangential and radial directions and f refers to the fact that it is caused by change in the fiber-saturation point m_f. If the assumption is made that longitudinal shrinkage is negligible, and that S_t and S_r are the tangential and radial moisture shrinkages, then expressions for $\alpha_{t,f}$ and $\alpha_{r,f}$ can be written

$$\alpha_{t,f} = G_0 \frac{S_t}{(S_t + S_r)} \frac{\Delta m_f}{\Delta T} \tag{4.51}$$

$$\alpha_{r,f} = G_0 \frac{S_r}{(S_t + S_r)} \frac{\Delta m_f}{\Delta T} \tag{4.52}$$

where $\Delta m_f / \Delta T$ is the rate of change of fiber-saturation point g/g per $°C$ in temperature. According to Stamm and Loughborough (1935) $\Delta m_f / \Delta T$ is approximately -0.001 (g/g)/°C. Using this value and taking S_t to be twice S_r, equations (4.51) and (4.52) reduce to

$$\alpha_{t,f} = -667\, G_0(10^{-6}) \tag{4.53}$$

$$\alpha_{r,f} = -333\, G_0(10^{-6}). \tag{4.54}$$

Table 4.7 shows some tabulated values of α_t and α_r calculated using equations (4.49) and (4.50) and also (4.53) and (4.54). It is clear that the latter effect dominates the dimensional behavior. Yokota and Tarkow (1962) measured a value for reversible thermal expansion α_t of -35×10^{-6} (°C)$^{-1}$ for Sitka spruce wood. This compares with a predicted value, using equations (4.49) and (4.53) of -228×10^{-6} (°C)$^{-1}$, assuming G_0 is 0.4 and m_f is 0.30 g/g. This disagreement in magnitude indicates that equations (4.53) and (4.54) predict a larger effect than has actually been found. However, the fact that the coefficient α_t was negative does show that this effect dominates the effect of normal thermal expansion. The curves obtained by Yokota and Tarkow are shown in Figure 4.21.

This same mechanism of changing fiber-saturation point with temperature also dominates dimensional changes with temperature below the freezing point of water at moisture contents above fiber saturation. However, in this case the effect gives a positive thermal expansion or an excess shrinkage when the wood cools below the freezing point of the water in the cell cavities. This effect, which Kübler (1962) calls coldness shrinkage, and its relationship to the change in

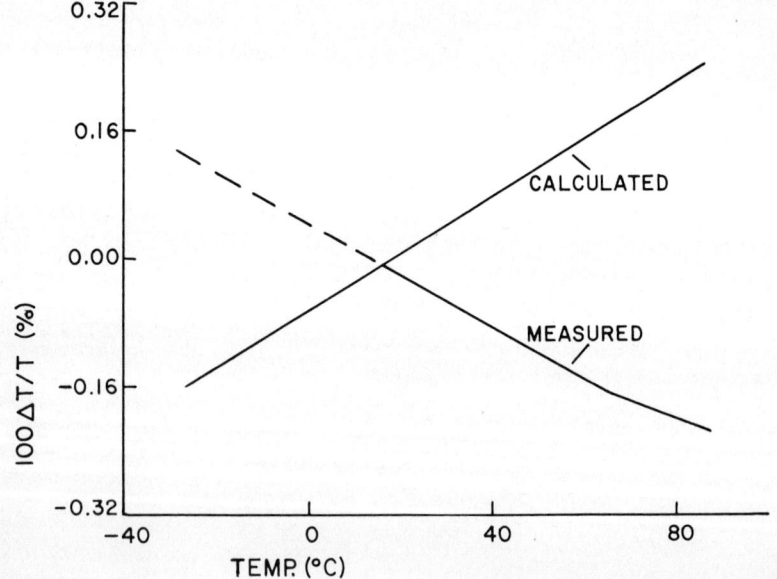

Fig. 4.21. Percent change in tangential dimensions (100 $\Delta T/T$) with temperature (°C) as calculated from normal thermal expansion and as measured by Yokota and Tarkow (1962).

Table 4.7. Values of α_t and α_r Calculated for Wood at Moisture Content Above m_f Where m_f Is Taken as 0.30 g/g

Dry Spec. Grav. G_0	Calculated expansion coefficient $(°C)^{-1}$ $(\times 10^6)$					
	$\alpha_{t,m}$ Equ. (4.49)	$\alpha_{t,f}$ Equ. (4.53)	$\alpha_{t,m} + \alpha_{t,f}$ Equ. (4.49) + (4.53)	$\alpha_{r,m}$ Equ. (4.50)	$\alpha_{r,f}$ Equ. (4.54)	$\alpha_{r,m} + \alpha_{r,f}$ Equ. (4.50) + (4.54)
0.2	21	-133	-112	15	-67	-52
0.4	39	-267	-228	29	-133	-104
0.6	56	-400	-344	43	-200	-157
0.8	72	-533	-461	56	-267	-211
1.0	87	-667	-580	68	-333	-265
1.2	101	-800	-699	81	-400	-319
1.4	114	-933	-819	92	-467	-375

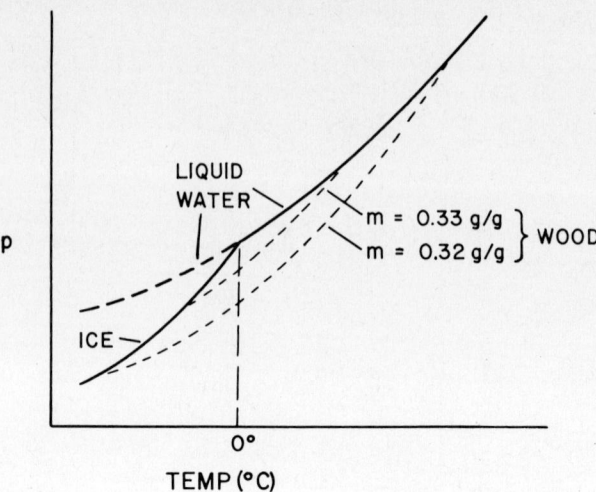

Fig. 4.22. Vapor pressure p (not to scale) against temperature of liquid water, ice, and wood near the fiber-saturation point m_f (g/g) in the vicinity of $0°C$ (adapted from Kübler 1962).

fiber-saturation point with temperature is explained by Kübler in terms of the vapor pressure-temperature diagram of Figure 4.22.

If moist wood is cooled from room temperature to the freezing point its fiber-saturation point m_f increases and the wood swells. This occurs because the vapor pressure p of the cell wall (Figure 4.22) at a given moisture content decreases more rapidly with decreasing temperature than does the saturated vapor pressure p_0 of liquid water. Therefore, if the cell wall is saturated at $m = m_f$ at room temperature its vapor pressure $p = p_0$. As temperature decreases, however, and p decreases more rapidly than p_0, then p becomes less than p_0 at constant m. The wood, therefore, is no longer saturated, and if it is still exposed to saturated vapor it will regain more moisture, thus increasing m_f and swelling in the process. It attains a maximum value of m_f at the freezing point of the water in the cell cavities.

As the wood is cooled below the freezing point, however, the reverse phenomenon takes place since the vapor pressure p_s of ice decreases more rapidly with decreasing temperature than does the vapor pressure p of the wood. Therefore the cell wall contains excess water which migrates out of the cell wall and condenses as ice in the cell cavity. The resulting shrinkage is much larger than the normal thermal shrinkage of the cell wall and is responsible for the coldness shrinkage of Kübler.

It is believed that coldness shrinkage is responsible for the formation of frost cracks in living trees as well as in green lumber subjected to excessively low temperatures. These frost cracks in living trees are a serious problem in some species (Schirp 1968) and cause a great deal of lumber degrade.

The third mechanism of thermal expansion is caused by the expansion of water when it freezes. This is important only when the cell cavities are nearly full of water so there is no room within the cell for the expansion to take place. Furthermore, it occurs only at the freezing point of the cell-cavity water which is usually slightly lower than that of free water because of the cell cavity capillaries and also because of the presence of water-soluble materials in the water. The fourth mechanism has already been described in sufficient detail for purposes of this text.

The irreversible thermal expansion mentioned earlier occurs when green wood is heated for the first time. It appears to be related to stresses that occur in the tree during growth and are relieved by heating in water. Kübler (1959) found that tangential expansions of one percent or more occurred in European beech and oak upon boiling in water. Yokota and Tarkow (1962) found a similar ir-reversible tangential expansion of up to nearly 0.5 percent in Sitka spruce sam-ples when heated to 90°C. However, they found a simultaneous irreversible radial contraction for the same heating conditions. These irreversible thermal dimensional changes are believed to be responsible for the large radial heart checks which often occur in veneer logs after heating in water or steam (MacLean 1952). Yokota and Tarkow state that the initial heating of green wood probably causes the breaking of hydrogen bonds which may be under stress in the living tree because of growth conditions. Figure 4.23 shows some of the dimensional

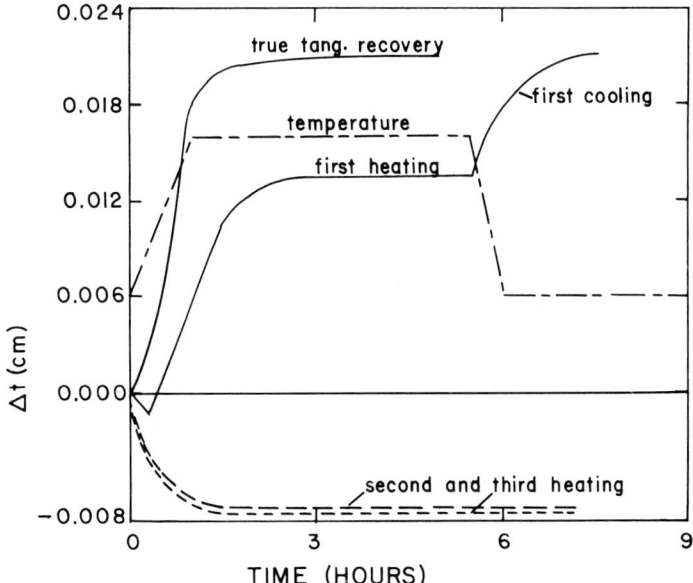

Fig. 4.23. Experimental curves of tangential dimensional changes Δt against time for green Sitka spruce heated and reheated in water according to the temperature schedule shown by broken line cycling between 25° and 90°C (adapted from Yokota and Tarkow 1962).

changes in green wood during the first heating and also during second and third heating periods.

Thermoelastic or Joule Effect

The thermoelastic effect, first studied by Joule (1859), is based on a theory developed by Lord Kelvin. It predicts that heat is generated when a mechanical compression stress is applied to an elastic solid such as wood. Application of a tensile stress should cool the specimen, and both effects are reversible if the material is perfectly elastic. The effect is related to the temperature changes associated with adiabatic expansion and contraction of gases and is analogous to the hygroelastic or Barkas effect described above. An equation of the form used by Joule will be derived here, following the treatment of Sinnott (1958).

When a solid is subjected to a hydrostatic pressure P its volume V decreases and work is done on the system. For the isothermal case at temperature T the increase dE in internal energy of the solid can be written as

$$dE = T(\partial P/\partial T)_v \, dV - P dV. \qquad (4.55)$$

However, the increase in internal energy dE is also equal to the difference in the heat increase dq in the solid, less the work done, or

$$dE = dq - P dV. \qquad (4.56)$$

Therefore, combining equations (4.55) and (4.56), and solving for dq/dV

$$dq/dV = T(\partial P/\partial T)_v. \qquad (4.57)$$

However, dq/dV is equal to $(\partial q/\partial T)_p \, (\partial T/\partial V)_p$, and therefore

$$(\partial q/\partial T)_p \, (\partial T/\partial V)_p = T(\partial P/\partial T)_v. \qquad (4.58)$$

But $(\partial q/\partial T)_p$ is equal to wc, where w is the mass of the solid and c its specific heat at constant pressure P. Also $(\partial V/\partial T)_p$ is equal to $\alpha_v V$, where α_v is the volume coefficient of thermal expansion. Making these substitutions into equation (4.58) and also substituting for w/V the term ρ for density, and rearranging, the equation obtained is

$$(\partial T/\partial P)_v = (\alpha_v T)/(\rho c). \qquad (4.59)$$

If the units of heat measurement and mechanical work are expressed in different units—if, for example, the specific heat c is in calories per gram-degree and pressure P is in dynes per square centimeters—then the conversion factor relating ergs to calories must be used in the equation. The volume change with stress in a solid is small, and the approximation may then be made that $(\partial T/\partial P)_v \approx dT/dP$, in which case equation (4.59) can be written

$$dT/dP \approx (\alpha_v T)/(J\rho c) \qquad (4.60)$$

where J is the mechanical equivalent of heat, say ergs per calorie.

In the case of a stress σ applied in one direction only, equation (4.60) may be written in the form used by Joule (1859), or

$$\Delta T \approx (\alpha T/(J\rho c))\sigma \qquad (4.61)$$

where α is the coefficient of linear expansion in the direction of application of stress. It should not be inferred that the Joule equation is an exact equation, because several simplifying assumptions have been made in its derivation. A similar statement could be made concerning the generalization of the Barkas equation (4.40) for a one-dimensional stress. A more rigorous treatment is complex and requires the use of elastic theory for anisotropic materials, which is outside the scope of this text.

Joule (1859) calculated from equation (4.61) that there should be an instantaneous decrease ΔT of $-0.023°K$ (or $°C$) when a parallel-to-grain tension load of 1,820 psi (125×10^6 dynes/cm^2) was applied to pine wood with a measured value of α, parallel to the grain, of $4 \times 10^{-6}(°K)^{-1}$. He measured a ΔT of $-0.017°K$ which demonstrated the correctness of the principle. Tests in compression parallel to the grain showed similar agreement with theory, although in this case ΔT was positive, as predicted. In compression tests perpendicular to the grain, Joule found an additional excess irreversible temperature rise which he correctly ascribed to the frictional or inelastic properties of the wood.

Recent corroboration of the Joule thermoelastic effect on wood has been obtained by Jimènez (1967), who applied a compressive load of 3,100 ±300 psi perpendicular to the grain to one cubical sample ($1.5 \times 1.5 \times 1.5$ inches) of each of seven high-density (air-dry densities 0.97 to 1.22 g/cc) tropical woods from Venezuela. He recorded the temperature in the interior of each sample as a function of time (Figure 4.24) and noted a mean temperature rise ΔT of ap-

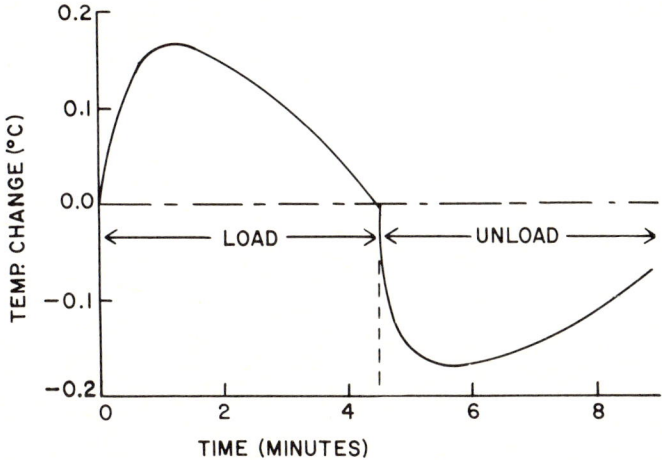

Fig. 4.24. Experimental curves of temperature against time observed by Jimènez (1967) for wood samples loaded transversely to 3,100 psi and then unloaded.

proximately $0.17°K$. Furthermore, when he unloaded the sample there was a temperature decrease of approximately the same amount, indicating the reversibility of the effect. Calculations based on the Joule equation for a wood density ρ of one g/cc and moisture content of 0.11 g/g (11 percent), the mean value for the samples used in this study, predict a temperature rise of $0.25°K$ immediately after application of load and a decrease of the same amount upon release of the load. This calculation is based on a mean temperature T of $300°K$, a transverse thermal-expansion coefficient α of 60×10^{-6} $(°K)^{-1}$, a specific heat c of 0.37 cal g^{-1} deg^{-1}, and a transverse stress σ of 3,100 psi $(21.4 \times 10^{7}$ dynes $cm^{-2})$.

Combined Hygroelastic and Thermoelastic Effects

It is clear from equations (4.44) and (4.61) that the thermoelastic effect is analogous to the hygroelastic effect; that is, an applied stress causes a change both in the temperature and in the moisture content independently of each other. There is also an interaction between these two effects, however. For example, the temperature increase associated with a compressive stress, as predicted by the thermoelastic equation, also raises the vapor pressure p of the wood. This is an additional increase in vapor pressure to that predicted by the hygroelastic equation alone. The reverse is true for a tensile stress.

Simpson (1970) observed the combination of these two effects in red oak samples which he subjected to compressive or tensile stresses applied in the radial and tangential directions in an atmosphere of constant temperature and humidity.

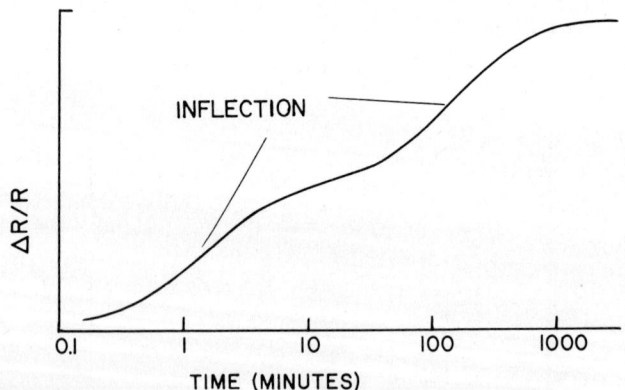

Fig. 4.25. Experimental curve of the form obtained by Simpson (1970), showing the increase in resistance R as a function of time after loading (logarithmic scale) for a compressive load.

For a compressive stress, for example, the electrical resistance of the sample decreased immediately after application of the load, presumably because of the thermoelastic heating effect since increasing temperature reduces the electrical resistance of wood (see Chapter 2). As time increased after loading, the sample cooled with a resultant increase in resistance with time. This occurred during the first few minutes after loading, with two-thirds of the change taking place in about two minutes. This is evidenced by the first inflection in the curve shown in Figure 4.25, in which time is on a logarithmic scale.

A second inflection occurring at about two hours after loading is also evident in Figure 4.25. It is believed to be caused by the diffusion of moisture out of the wood as it comes to equilibrium with the stressed condition of the wood at a lower moisture content than the initial value in the stress-free condition, because of the hygroelastic effect. It is known that moisture diffuses much more slowly than heat in wood (Siau 1971), in this case approximately sixty times more slowly. Since these were samples 0.5 by 0.5 inches in cross-section with the longitudinal structural direction as one of these short dimensions, most of the moisture and heat flow was undoubtedly parallel to the grain direction.

5. Theories of Water Sorption by Wood

The general nature of water sorption by wood has already been discussed in Chapter 2, and the thermodynamics of sorption were examined in Chapter 4. We will now consider some of the more quantitative theories which have been used to explain the mechanism of water sorption in wood.

Many theories have been proposed to explain the sorption of water by hygroscopic polymers. For textile materials these fall into two categories, according to King in Hearle and Peters (1960). One category, exemplified by the BET equation (Brunauer, Emmett, and Teller 1938) considers the process to be one of sorption of water on internal sorption sites or surfaces in the material. The second category, represented by the Hailwood and Horrobin (1946) theory, considers the process of sorption to be one of solution in which there may be two or more phases at equilibrium. A third category, represented by the deformation-in-space theory of Malmquist (1958), makes no assumptions concerning the mechanism of sorption but considers the process in terms of mass and space relationships within the cell wall of wood as functions of wood moisture content and vapor pressure. It also takes into account the cohesive strength or resistance to deformation of the cell wall.

Venkateswaran (1970) lists eighteen sorption isotherm equations which have been proposed to explain the mechanism of sorption on cellulosic materials. He points out that in the derivation of most of these equations, one or more common assumptions are made. In terms of water sorption by wood these assumptions can be stated as follows: the energy of interaction with the dry wood of the first or primary water molecules taken up is generally higher than the interaction energy of secondary water molecules taken up by the more moist material; the interaction energy of the secondary water molecules with the moist wood is essentially equal to the heat of condensation of liquid water and is constant; there is no energy of interaction between adjacent sorbed water molecules. Some of the sorption equations are derived from classical thermodynamic considerations, other from statistical considerations, and others from combinations of these or other treatments.

We shall not attempt to derive all of these equations here but will derive representative equations, including the BET equation, the Hailwood and Horrobin equation, and the Malmquist deformation-in-space equation. We shall give some

of the other equations in final form and shall apply a number of them to available sorption data, particularly the experimental data published in the *Wood
Handbook* (1955) of the United States Forest Products Laboratory.

Surface Sorption Theory of Brunauer, Emmet, and Teller (BET Theory)

There are several theories of moisture sorption which postulate that moisture
is taken up by wood on internal surfaces which are presumed to preexist in the
dry wood. The most popular of these theories is the BET theory which proposes
that one-to-many layers of water molecules can be sorbed on the internal surfaces of sorptive materials such as wood. The BET theory is based on an earlier
theory of Langmuir who assumed that a single layer of vapor molecules could be
sorbed as a monolayer on these internal surfaces. We will first review this earlier
theory of Langmuir's since it provides a basis for the later BET theory.

Langmuir Model

Langmuir (1918) considered that a surface which is capable of adsorbing a
vapor from the atmosphere could be thought of as containing two parts at any
particular vapor pressure. Langmuir was not thinking of wood particularly when
he derived his theory, but we shall speak in terms of wood and water vapor,
keeping in mind that this is a general relationship between the vapor pressure of
any gas and the amount of the gas adsorbed on a surface of any kind.

Referring to Figure 5.1, we can consider that the wood surface A_0 is that por-

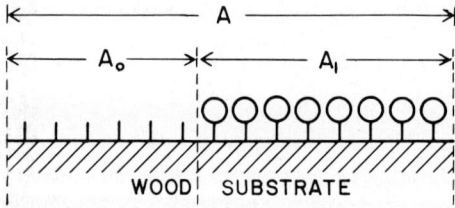

Fig. 5.1. Langmuir sorption model, showing the surface A_0 not covered by water molecules and the surface A_1 covered by a monolayer.

tion of the total internal surface A which is not covered by a layer of water
molecules. The rest of the surface of area A_1 is covered completely with a
monolayer of water.

When the wood surface is at equilibrium with the moisture in the air, the rate
of evaporation of water from the surface A_1 is equal to the rate of condensation
on the surface A_0. The assumption is now made that the rate of evaporation is

proportional to the area A_1. Thus, the rate of evaporation $= k_1 A_1$, where k_1 is a constant of proportionality.

Likewise, the rate of condensation is assumed to be proportional to the area A_0, with the proportionality constant k_0. Thus, the rate of condensation $= k_0 A_0$. The constant k_0 is presumed to be proportional to the vapor pressure p of the water vapor near the wood surface, or $k_0 = ap$ where a is a constant relating k_0 and p.

At equilibrium $k_1 A_1 = k_0 A_0$, or

$$k_1 A_1 = ap A_0 = ap(A - A_1) \qquad (5.1)$$

and

$$p = \frac{k_1 A_1}{a(A - A_1)} = \left(\frac{k_1}{a}\right)\left(\frac{A_1/A}{1 - A_1/A}\right). \qquad (5.2)$$

If it is now assumed that the wood moisture content M is proportional to A_1, and that the moisture content at fiber saturation M_f is proportional to the total area A, then equation (5.2) reduces to

$$p = k\left(\frac{M/M_f}{1 - M/M_f}\right); \quad \text{or} \quad \frac{M}{M_f} = \frac{p}{p + k} = \frac{h}{k/p_0 + h} \qquad (5.3)$$

where $k = k_1/a$ and $h = p/p_0$. Equation (5.3) predicts a sorption isotherm of the form shown in Figure 5.2 and is characteristic of monolayered surface sorption.

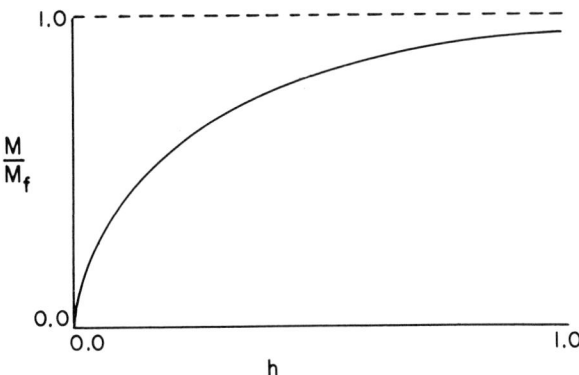

Fig. 5.2. Langmuir sorption isotherm, plotted as the ratio M/M_f, where M_f is the saturation moisture content, against relative vapor pressure h.

BET Multilayer Model

The BET theory is an extension of the Langmuir theory. It considers that there may be a number of layers of water sorbed onto the internal surfaces of

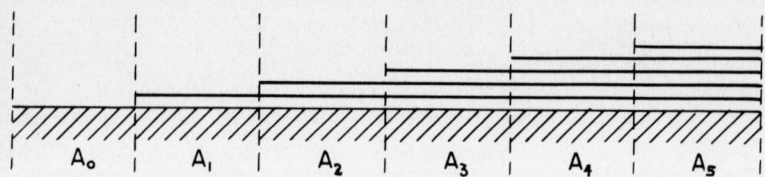

Fig. 5.3. BET sorption model, showing the surfaces A_0, A_1, A_2, etc., covered by 0, 1, 2, etc., layers of water molecules.

materials such as wood. Consider, for example, Figure 5.3, which shows schematically several layers of water covering the internal surface of wood. A certain area A_0 of the wood is assumed to be free of any water, as in the Langmuir model. The area A_1 is covered by one layer, A_2 by two layers, A_3 by three layers, and so on.

As in the case of Langmuir's monolayer theory, there are assumed to exist equilibria between the rates of condensation and of evaporation of adjacent areas. For example,

$$\text{for layer no. 1,} \qquad a_1 p A_0 = k_1 A_1 \qquad\qquad (5.4a)$$

$$\text{for layer no. 2,} \qquad a_2 p A_1 = k_2 A_2 \qquad\qquad (5.4b)$$

$$\text{for layer no. 3,} \qquad a_3 p A_2 = k_3 A_3 \qquad\qquad (5.4c)$$

$$\text{for layer no. } i, \qquad a_i p A_{i-1} = k_i A_i \qquad\qquad (5.4d)$$

The first term in each case is the rate of condensation on area A_{i-1}, and the second term is the rate of evaporation from the area A_i.

The coefficients k_1, k_2, k_3, ..., k_i, are proportional to the fraction of molecules with sufficient energy to evaporate from the particular layer of water. If it is assumed that the Boltzmann distribution applies (see Chapter 1), they may be defined in terms of the energies U_1, U_2, U_3, ..., U_i required to evaporate a mole of water from the particular layer. Thus,

$$k_1 = b_1 \exp(-U_1/RT) \qquad\qquad (5.5a)$$

$$k_2 = b_2 \exp(-U_2/RT) \qquad\qquad (5.5b)$$

$$k_i = b_i \exp(-U_i/RT) \qquad\qquad (5.5c)$$

where b_1, b_2, ..., b_i, are constants which are characteristic of the particular layer $(1, 2, \ldots, i)$.

The total internal wood surface, designated by A, is given by

$$A = \sum_{i=0}^{\infty} A_i = (A_0 + A_1 + A_2 + A_3 + \cdots + A_i) \qquad\qquad (5.6)$$

The total volume of water V_m possible in the monolayer is given by

$$V_m = v_0 \sum_{i=0}^{\infty} A_i = v_0(A_0 + A_1 + A_2 + \cdots + A_i) \tag{5.7}$$

where v_0 is defined as the volume of water per unit layer of water per unit of area of internal wood surface. Thus, the total volume V of water in all the layers in the wood is

$$V = v_0 \sum_{i=0}^{\infty} i.A_i = v_0(1.a_1 + 2.A_2 + \cdots + i.A_i) \tag{5.8}$$

The ratio V/V_m from equations (5.7) and (5.8) is

$$\frac{V}{V_m} = \frac{(A_1 + 2A_2 + 3A_3 + 4A_4 + \cdots + iA_i)}{(A_0 + A_1 + A_2 + A_3 + A_4 + \cdots + A_i)} . \tag{5.9}$$

If we know that the ratio of any two successive areas is the same as that of any other pair of successive areas (except for A_1 and A_0), and designate this ratio as x, then when $i \neq 1, x$ is defined as

$$x = A_i/A_{i-1} = A_2/A_1 = A_3/A_2 = A_4/A_3 \tag{5.10}$$

or

$$A_i = A_{i-1}x; A_2 = A_1 x; A_3 = A_2 x; \text{etc.} \tag{5.11}$$

and substitute these relationships into equation (5.9) we find

$$\frac{V}{V_m} = \frac{(A_1 + 2xA_1 + 3x^2A_1 + 4x^3A_1 + \cdots)}{(A_0 + A_1 + xA_1 + x^2A_1 + x^3A_1 + \cdots)} \tag{5.12}$$

or

$$\frac{V}{V_m} = \frac{A_1(1 + 2x + 3x^2 + 4x^3 + \cdots)}{A_0 + A_1(1 + x + x^2 + x^3 + \cdots)} \tag{5.13}$$

The binomial theorem states that

$$(1 - x)^{-2} = (1 + 2x + 3x^2 + 4x^3 + \cdots) \tag{5.14a}$$

and

$$(1 - x)^{-1} = (1 + x + x^2 + x^3 + \cdots) \tag{5.14b}$$

When equations (5.14a) and (5.14b) are substituted into (5.13)

$$\frac{V}{V_m} = \frac{A_1(1 - x)^{-2}}{A_0 + A_1(1 - x)^{-1}} . \tag{5.15}$$

The relationship between the area A_0 and A_1 is taken to be the same as in the case of the Langmuir isotherm given by equation (5.1) or equation (5.4a). Thus

$$A_1/A_0 = a_1 p/k_1 \qquad (5.16)$$

or, using the definition of k_1 given in equation (5.5a)

$$A_1/A_0 = (a_1 p/b_1)\exp(U_1/RT). \qquad (5.17)$$

If we now assume that there is a constant relationship between the ratio A_1/A_0 and A_i/A_{i-1} for $i \neq 1$, then we can write in terms of the constant C, relating these ratios, that

$$A_1/A_0 = C(A_i/A_{i-1})_{i \neq 1} = Cx \qquad (5.18)$$

since x is defined in this way by equation (5.10). Thus

$$A_1 = A_0 Cx \qquad (5.19)$$

which, when substituted into equation (5.15), reduces it to

$$\frac{V}{V_m} = \frac{A_0 Cx(1-x)^{-2}}{A_0 + A_0 Cx(1-x)^{-1}} = \frac{Cx}{(1-x+Cx)(1-x)}. \qquad (5.20)$$

According to the original BET theory, there is no limit to the number of layers of water which the wood can sustain. Therefore, the volume V and the ratio V/V_m approach infinity as the wood becomes saturated. According to equation (5.20) this happens as x approaches unity. Thus

$$\frac{V}{V_m} \longrightarrow \infty \quad \text{as} \quad x \longrightarrow 1. \qquad (5.21)$$

Therefore, from the definition of x given by equation (5.10), combined with equation (5.4d) and (5.5c)

$$x = (A_i/A_{i-1})_{i \neq 1} = a_i p/k_i = (a_i p/b_i)\exp(U_i/RT) \qquad (5.22)$$

when $x = 1$, p must $= p_0$, the saturated vapor pressure, and

$$1 = p_0(a_i/b_i)\exp(U_i/RT). \qquad (5.23)$$

Dividing equation (5.22) by (5.23) gives the general relationship that $x = p/p_0 = h$, the relative vapor pressure, because

$$\frac{x}{1} = \frac{(a_i/b_i)p\exp(U_i/RT)}{(a_i/b_i)p_0\exp(U_i/RT)} = p/p_0 = h. \qquad (5.24)$$

If we now substitute for x in equation (5.20) $h = p/p_0$, it then becomes

$$\frac{V}{V_m} = \frac{Ch}{(1-h+Ch)(1-h)} \qquad (5.25)$$

which can be rearranged to give the standard form of the BET equation. Thus

$$\frac{h}{V(1-h)} = \frac{1}{V_m C} = \frac{(C-1)h}{V_m C}. \qquad (5.26)$$

The BET equation can also be written in terms of moisture content M if it is assumed that M is proportional to the volume V of water in the wood. With this assumption equation (5.26) becomes

$$\frac{h}{M(1-h)} = \frac{1}{M_m C} + \frac{(C-1)h}{M_m C}. \qquad (5.27)$$

Equation (5.27) predicts that a straight-line relationship exists between the ratio $h/M(1-h)$ and the relative vapor pressure h, as shown in Figure 5.4, and

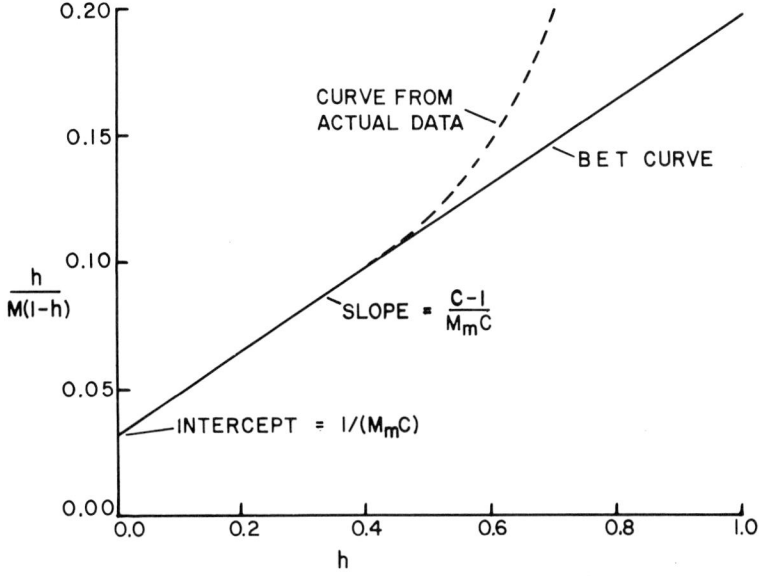

Fig. 5.4. BET isotherm plot on data at $110°F$ from *Wood Handbook* showing deviation from the BET curve at higher relative vapor pressures.

that the slope of this line is equal to $(C-1)/(M_m C)$, with the zero intercept equal to $1/(M_m C)$. Thus, both the constants M_m and C can be calculated if the theoretical relationship is obeyed.

The constant M_m is interpreted to be the moisture content of the wood corresponding to complete monomolecular sorption. It does not mean that this monolayer is completely filled before any of the additional layers form, although a particular sorption site must have a monolayer before subsequent layers can be formed.

The constant C was defined by equation (5.18) as the ratio of A_1/A_0 to A_i/A_{i-1}, for $i \neq 1$. If we combine equation (5.18) with equations (5.4a) and (5.4d), it becomes

$$C = \frac{(a_1)(k_i)}{(k_1)(a_i)_{i \neq 1}} \qquad (5.28)$$

which, when k_1 and k_i as defined in equations (5.5a) and (5.5c) are substituted, becomes

$$C = (a_1/b_1)(b_i/a_i)\exp\left[(U_1 - U_i)/RT\right]. \qquad (5.29)$$

If now the product $(a_1/b_1)(b_i/a_i)$ is designated by a constant B, and if U_i, the energy required to evaporate a mole of water from any layer but the first, is taken to be equal to the molar heat of vaporization U_0 of free water $(= 18\,Q_0$, see Chapters 1 and 4), C can be defined as

$$C = B \exp\left((U_1 - U_0)/RT\right). \qquad (5.30)$$

The difference in the energies U_1 and U_0 should be equal to $18\,Q_L$, if the assumption is made that all of the heat generated during sorption of moisture by wood from the liquid state is caused by the sorption of the monolayer. Furthermore, it is expected to be equal to the value of Q_L at zero moisture (about 280 calories per gram of water). According to Stamm (1964), the constant B in equation (5.30) is close to unity.

The BET equation fits the sorption isotherm for wood reasonably well at low moisture contents, up to a relative vapor pressure of about 0.4. Above this point the deviation from linearity of the data as plotted in Figure 5.4 becomes quite large and no longer fits the simple theory. In practice the linear portion of the curve (below about $h = 0.4$) is used to calculate M_m and C. It is believed that the deviations of the actual curve from the simple BET straight-line curve is caused by the assumption in the original BET theory that an infinite number of layers of water may form on the internal surfaces. This is obviously an impossibility and Brunauer, Emmett, and Teller (1938) modified their equation so as to include a restriction on the maximum number of layers which may form on the surface. The modified form of the equation can be written as,

$$\frac{h}{M(1-h)} = \frac{1}{M_m C}\,\frac{1 + (C-1)h - Ch^{n+1}}{1 - (n+1)h^n + nh^{n+1}} \qquad (5.31)$$

where all the terms h, M, M_m, and C have the same meanings as in equation (5.27), and the term n represents the maximum number of layers of water permitted on any sorption site.

The BET equations can also be derived by statistical methods such as are given by Hill (1946) (see Hearle and Peters 1960) and by Rounsley (1961) (see Venkateswaran 1971).

It can be shown that equation (5.31) reduces to the unmodified BET equation (5.27) when $n = \infty$. This is true because the term h^{n+1} approaches zero as n approaches infinity, since h is less than unity. Also $(n + 1)h^n$ approaches nh^{n+1} as n approaches infinity, and the two terms cancel each other, leaving only $M_m C$ in the denominator and $1 + (C - 1)h$ in the numerator, identical with equation (5.27).

It can also be shown that equation (5.31) reduces to the simple Langmuir isotherm of equation (5.3) when $n = 1$. When $n = 1$, equation (5.31) becomes

$$\frac{h}{M(1 - h)} = \frac{1}{M_m C} \frac{1 + (C - 1)h - Ch^2}{1 - 2h + h^2}$$

which reduces, after rearrangement and cancellation of terms, to

$$\frac{M}{M_m} = \frac{Ch}{1 + Ch} = \frac{h}{(1/C) + h} \qquad (5.32)$$

identical with the Langmuir equation (5.3) if $C = p_0/k$ and $M_m = M_f$. From this relationship it can be seen that the constant k in equation (5.3) really is equal to (k_1/a_i), since

$$k = \frac{p_0}{C} = \frac{(b_i/a_i)\exp(-U_i/RT)}{(a_1/b_1)(b_i/a_i)\exp[(U_1 - U_i)/RT]} \qquad (5.33)$$

by using p_0 from equation (5.23) and C from equation (5.28). This reduces to

$$k = k_1/a_1 \qquad (5.34)$$

from equation (5.5a).

We will now return to our discussion of the modified BET equation (5.31). In order to use this equation it is first necessary to evaluate the constants C and M_m by use of the linear portion of the originally unmodified BET equation (5.27), as shown in Figure 5.4. This procedure introduces the assumption that there is little difference between the modified and unmodified BET equations at low humidities. These values are then substituted into equation (5.31), together with various values of n and of relative vapor pressure h, to obtain the theoretical BET sorption isotherms for different values of n. It is most convenient for this purpose to rewrite equation (5.31) into the form

$$M = \frac{M_m [Ch(1 - (n - 1)h^n + nh^{n+1})]}{(1 - h)[1 - (C - 1)h - Ch^{n+1}]} . \qquad (5.35)$$

Figure 5.5 shows the form of such curves for different values of n, together with the original sorption isotherm from which the constants M_m and C were evaluated. It can be seen that a better fit to the actual experimental isotherm, shown by the solid curve, is obtained at higher values of h when $n = 6, 7$, or 8 than when $n = \infty$. This is interpreted to mean that a maximum of six, seven, or eight

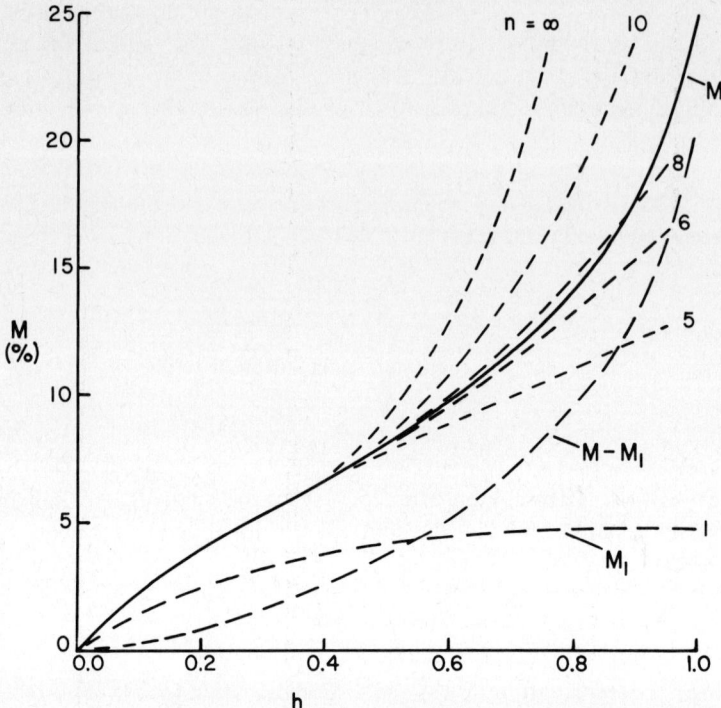

Fig. 5.5. Comparison of BET sorption isotherms (broken lines) for various values of n with the observed isotherm using same data as in Figure 5.4. Also shown is the difference $M - M_1$ between the observed curve and the monolayer or Langmuir curve, M_1.

layers of water can cover a particular sorption site. The average number of layers at any particular value of h can be estimated by taking the ratio of M, the total moisture content, to the value M_1, corresponding to the Langmuir isotherm (the lowest curve in Figure 5.5). Thus from Figure 5.5 the average number of layers, or of molecules per sorption site, M/M_1, are approximately 1.5, 1.8, 2.5, and 3.5 at $h = 0.3, 0.5, 0.7$, and 0.9, respectively.

Our next comments on the BET theory relate to the values found for the constant C. According to equation (5.30) the following relationship holds

$$U_1 - U_0 = RT \ln(C/B) \qquad (5.36)$$

or

$$Q_L = (RT/18) \ln(C/B) \qquad (5.37)$$

since the difference $(U_1 - U_0)$ should equal $18 Q_L$, as indicated previously. If B is taken as unity, as is suggested on page 140 of Stamm (1964), equation

(5.37) can be evaluated using $C = 6.3$ from Figure 5.4. Thus, at $110°F$, $T = 273.1 + 43.3° = 316.4°K$, and

$$Q_L = \frac{(1.987)(316.4)\ln(6.3)}{18} = 64.4 \text{ cal/g-water.} \tag{5.38}$$

This is considerably lower than the value of Q_L of 260–280 calories per gram of water obtained for dry wood. However, it is close to the mean value of Q_L over the entire hygroscopic range.

The modified BET equation has been fitted by Simpson (1971, personal communication) to the sorption data given in the *Wood Handbook* at individual temperatures over the temperature range from $30°F$ to $210°F$. The results indicate that the constant C is invariant with temperature, having a value 5.925 ± 0.732. This gives a mean value of Q_L, obtained from equation (3.37) of 62.2 calories per gram of water. The constant M_m in the BET equation decreases with increasing temperature according to Simpson's results as

$$M_m = 7.4 - 0.020\,F \quad (\%) \tag{5.39}$$

and the term n increases as

$$n = 4.6 + 0.022\,F \tag{5.40}$$

where F is the temperature in Fahrenheit degrees.

The interpretation of the above results is that within the limits of reliability of the sorption data, the maximum moisture content M_m of the monolayer, as defined in the BET model, decreases with increasing temperature. This is as anticipated since the number of available sorption sites is expected to decrease with increasing temperature. There is no ready explanation of the increase in n with increasing temperature, however.

Solution Theory of Hailwood and Horrobin

The Hailwood and Horrobin (1946) sorption theory was derived specifically to explain the sigmoid shape of the sorption isotherm of water vapor on polymers, particularly textile polymers. It has also been applied in recent years to the sorption isotherm for wood. It has been criticized for some of the assumptions used in its derivation, but it appears to yield at least certain fundamental parameters which give a reasonable description of the sorption process for wood, and it will be derived here.

The theory of Hailwood and Horrobin considers that water sorbed by wood exists partly as water of hydration and partly in solid solution. The three chemical species are polymer (dry cellulose or wood in our case), hydrated polymer (hydrated cellulose or "chemisorbed" water), and dissolved water. They are considered to behave as an ideal solution.

It is most convenient to treat this theory in terms of molar concentrations. We shall let X_0 symbolize the number of moles of unhydrated or dry wood, X_h the number of moles of hydrated wood (and therefore also the number of moles of water of hydration if we assume one mole of water of hydration per mole of hydrated wood), and X_s the number of moles of dissolved or unhydrated water. We do not know what the molecular weight, W, of the wood is at this point, but according to the theory we should be able to calculate it in terms of the molecular weight per mole of water sorption sites. This latter term actually means that the wood is presumed to have sorption sites to which water molecules may attach themselves and thus become water of hydration. If we know the number of moles of sorption sites per gram of dry wood we can calculate the molecular weight per sorption site based on the Hailwood and Horrobin theory.

The total number of moles of the three basic species is $X_h + X_0 + X_s$. We now assume that this solution behaves as an ideal solution; that is, the activities A_h, A_0, and A_s, of the hydrated water, dry wood, and dissolved water are equal to their mole fractions in the solution. In other words

$$A_h = \frac{X_h}{(X_h + X_0 + X_s)} \tag{5.41a}$$

$$A_0 = \frac{X_0}{(X_h + X_0 + X_s)} \tag{5.41b}$$

$$A_s = \frac{X_s}{(X_h + X_0 + X_s)}. \tag{5.41c}$$

When an equilibrium condition exists in a system consisting of the three components, the equilibrium constant, designated here as K_1^*, is defined as the ratio of the chemical activity A_h of the reaction product (hydrate) to the product of the activities of the two reactants (water and unhydrated wood), $A_0 A_s$. Thus, using equation (5.41)

$$K_1 = A_h / A_0 A_s = X_h / X_0 A_s, \quad \text{or} \quad X_h = K_1 X_0 A_s. \tag{5.42}$$

There also exists an equilibrium between the dissolved water and its relative vapor pressure, p/p_0 or h. The equilibrium constant $K_2^\dagger$ for this system is given as

$$K_2 = A_s / h; \quad \text{or} \quad A_s = K_2 h. \tag{5.43}$$

---

*The reaction of water and dry wood can be written dissolved water (activity A_s) + dry wood (activity A_0) $\overset{K_1}{\rightleftarrows}$ hydrated wood (activity A_h) where K_1 is the equilibrium constant.

†The equilibrium condition between water vapor and dissolved water can be written water vapor (activity p/p_0) $\overset{K_2}{\rightleftarrows}$ dissolved water (activity A_s) where K_2 is the equilibrium constant.

The ratio $X_h/(X_h + X_0)$ gives the moles X_h of hydrated wood and also of water of hydration per mole of hydrated plus unhydrated wood or per mole of dry wood. Thus, combining equations (5.42) and (5.43) gives for $X_h/(X_h + X_0)$

$$\frac{X_h}{X_h + X_0} = \frac{K_1 X_0 K_2 h}{K_1 X_0 K_2 h + X_0} = \frac{K_1 K_2 h}{K_1 K_2 h + 1} \tag{5.44}$$

We also would like to calculate the ratio $X_s/(X_h + X_0)$, which gives the moles of dissolved water per mole of dry wood. To do this, we write equation (5.41c) in inverted form and rearrange it to find

$$(X_h + X_0)/X_s = (1/A_s) - 1 = (1 - A_s)/A_s \tag{5.45}$$

and, inverting, using equation (5.43) to eliminate A_s, we find

$$\frac{X_s}{X_h + X_0} = \frac{K_2 h}{(1 - K_2 h)} . \tag{5.46}$$

The sum of equations (5.44) and (5.46) gives the total moles of water in the wood per mole of dry wood. This can be related to wood moisture content M by recalling that the moles of water are equal to the grams of water divided by the molecular weight of water (18 grams per mole), and that the moles of dry wood are equal to the grams of dry wood divided by the molecular weight of the wood per mole of sorption sites. The latter is in general unknown, and will be represented by the symbol W. Thus,

$$\frac{X_h}{X_h + X_0} + \frac{X_s}{X_h + X_0} = \left[\frac{(\text{g hydrated water}/18) + (\text{g dissolved water}/18)}{\text{g dry wood}/W} \right] \tag{5.47}$$

$$\frac{X_h + X_s}{X_h + X_0} = \frac{W}{18} \left(\frac{\text{g hydrated water}}{\text{g dry wood}} + \frac{\text{g dissolved water}}{\text{g dry wood}} \right) \tag{5.48}$$

$$\frac{X_h + X_s}{X_h + X_0} = (W/18)(m_h + m_s) = (W/18)m \tag{5.49}$$

where m_h and m_s are the equivalent fractional moisture contents of the hydrated and dissolved water and m is the total fractional moisture content, all based on dry weight of the wood.

Equation (5.49) can be combined with equations (5.44) and (5.46) to give

$$m = m_h + m_s = \frac{18}{W} \left(\frac{K_1 K_2 h}{1 + K_1 K_2 h} + \frac{K_2 h}{1 - K_2 h} \right) \tag{5.50}$$

or, in terms of percent relative humidity, $H = 100\,h$, and percent moisture content, $M = 100\,m$

$$M = M_h + M_s = \frac{1{,}800}{W} \left(\frac{K_1 K_2 H}{100 + K_1 K_2 H} \right) + \frac{1{,}800}{W} \left(\frac{K_2 H}{100 - K_2 H} \right) . \tag{5.51}$$

The first term on the right is equivalent to M_h, the percent moisture content consisting of water of hydration. The second term is equivalent to M_s, the percent moisture content consisting of water of solution or dissolved water.

The term 18/W in equations (5.50) or (5.51) is equal to the ratio of the grams of water per mole of water in the wood, 18, to the grams of dry wood per mole of sorption sites, W. It is therefore equivalent to the fractional moisture content m_1 of the wood when there is one molecule of water on each sorption site—that is, when each sorption site is hydrated with one water molecule—since in this case there is one mole of water per mole of sorption sites. Therefore

$$m_1 = 18/W \text{ (g/g)}; M_1 = 1{,}800/W \text{ (\%)}. \tag{5.52}$$

According to equations (5.50) or (5.51) there are three constants K_1, K_2, and W (or m_1) which determine what the relationship of M and H (the sorption isotherm) will be. Furthermore, these three constants also determine what portion of total moisture content M at any given humidity H is water of hydration M_h and what portion is water of solution M_s.

It is also possible, by use of equation (5.50) or (5.51), to calculate the constants K_1, K_2, and W if a sorption isotherm relating M and H is known. These calculations are tedious with the equations in the form shown, but it is possible to transform these equations into a simpler form which can be evaluated more easily. This can be performed with equation (5.51), for example, by taking the common factor H from the right to the left side and then inverting the equation to solve for H/M, thus

$$\frac{H}{M} = \frac{W}{1{,}800}\left[\frac{(100 + K_1 K_2 H)(100 - K_2 H)}{K_1 K_2 (100 - K_2 H) + K_2 (100 + K_1 K_2 H)}\right]$$

$$\frac{H}{M} = \frac{W}{18}\left[\frac{1}{K_2 (K_1 + 1)}\right] + \frac{W(K_1 - 1)H}{1{,}800 (K_1 + 1)} - \frac{W}{180{,}000}\left(\frac{K_1 K_2 H^2}{K_1 + 1}\right) \tag{5.53}$$

or

$$\frac{H}{M} = A + BH - CH^2 \tag{5.54}$$

where

$$A = \frac{W}{18}\left[\frac{1}{K_2 (K_1 + 1)}\right]; B = \left(\frac{W}{1{,}800}\right)\left(\frac{K_1 - 1}{K_1 + 1}\right); C = \left(\frac{W}{180{,}000}\right)\left(\frac{K_1 K_2}{K_1 + 1}\right). \tag{5.55}$$

From equations (5.54) and (5.55) it is seen that the Hailwood and Horrobin theory predicts a parabolic relation between the ratio H/M and the relative humidity H, since A, B, and C are constants related to the fundamental constants

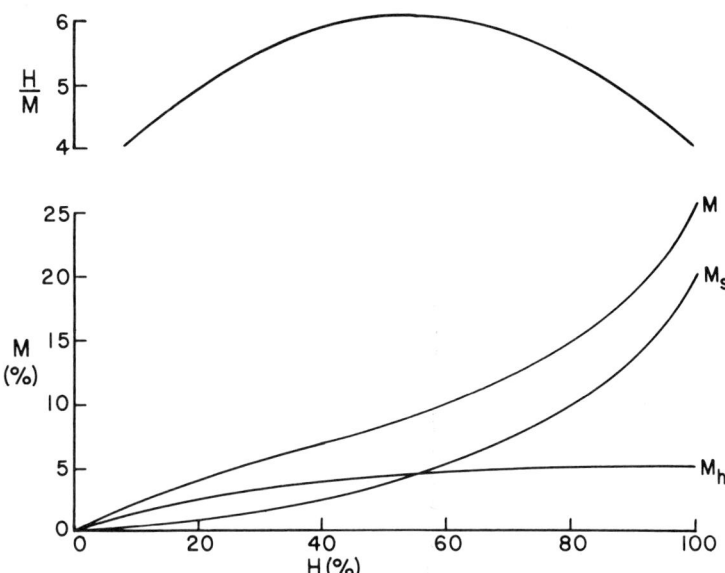

Fig. 5.6. Sorption isotherm for wood at $110°F$ from data in *Wood Handbook* including Hailwood and Horrobin curves of M, M_h, and M_s (%) as functions of relative humidity H (%). Also shown is the fitted curve of H/M against H.

W, K_1 and K_2. The constants A, B, and C can be evaluated empirically by fitting a parabolic equation to the sorption isotherm plotted in the form H/M versus H.

Figure 5.6 shows the curve of best fit of H/M against H for the sorption data at $110°F$ given in the *Wood Handbook*. Also shown are the calculated curves of M_h, M_s, and their sum M, all plotted as functions of humidity H.

Spalt (1958) first applied the Hailwood and Horrobin theory to sorption isotherms of water vapor on wood. He calculated A, B, and C, as well as K_1, K_2, and W for eight softwoods and eight hardwoods at a temperature of $90°F$, using both desorption and adsorption curves. Some of his data as well as the values of M_1 are shown in Table 5.1, including the data for acetylated white spruce; that is, wood which has had some of its hydroxyl groups replaced with acetyl groups to reduce its hygroscopicity and thereby increase its dimensional stability.

There are several interesting points to observe in the data of Table 5.1. First, the value for $M_1 = 1{,}800/W$ is always lower for adsorption than for desorption, indicating that there are fewer sorption sites available for adsorption than for desorption. This seems reasonable because after wood is dried some of the sorption sites undoubtedly crosslink and are not available for sorption during subsequent adsorption. After it is rewetted and the desorption isotherm is measured, these sites are opened up by the swelling forces of the water and the number of sites increases.

Table 5.1. Hailwood and Horrobin Coefficients for Wood at 90°F (Spalt 1958)

Kind of Wood	A	B	C	K_2	K_1	W	1,800/W*
Redwood (Des.)	1.774	0.0985	0.000725	0.659	9.43	219	8.22
Redwood (Ads.)	2.267	0.1384	0.001172	0.754	9.09	311	5.79
Basswood (Des.)	2.800	0.0884	0.000830	0.758	5.17	236	7.63
Basswood (Ads.)	2.321	0.1637	0.001520	0.830	9.49	364	7.49
W. Spruce (Des.)	1.388	0.1030	0.000839	0.741	11.02	222	8.11
W. Spruce (Ads.)	2.209	0.1213	0.001033	0.750	8.32	278	6.47
Acetyl. W. Spruce (Des.)	10.377	0.1320	0.001466	0.712	2.79	504	3.57
Acetyl. W. Spruce (Ads.)	11.744	0.1969	0.002068	0.731	3.29	664	2.71

*Value of 1,800/W is equivalent to $M_1(\%)$ by equation (5.52).

A second observation is that there is considerable variation among the woods with respect to the values for the constants for either adsorption or desorption. This indicates substantial differences in hygroscopicity among the woods.

A third point of interest is to note the strong decrease in the value of $1,800/W = M_1$ when white spruce is acetylated. This indicates a large reduction in the number of sorption sites available in the wood after acetylation, which is as anticipated because the process of acetylation replaces some of the hydroxyl groups in the wood with acetyl groups. The hydroxyl groups are believed to be the primary sorption sites for water in wood, and replacing these with the much less hygroscopic acetyl group is expected to reduce moisture sorption.

Similar effects have been noted when water-soluble extractives are removed from wood. For example, Wangaard and Granados (1967) found a decrease in W (or increase in M_1) in many cases after woods with high extractive contents were extracted with water. This is interpreted as meaning that the removal of extractives makes available additional moisture sorption sites.

It is known that increasing the temperature reduces the hygroscopicity of wood. That is, for a given relative humidity the equilibrium moisture content of wood is lower as temperature increases. This is a thermodynamic necessity if the value of Q_L as obtained from the sorption isotherms by use of the Clausius-Clapeyron equation is to be greater than zero (see Chapter 4). Since increasing the temperature does reduce the hygroscopicity, it is not surprising that W increases with temperature. For example, Simpson (1971) fitted the sorption data given in the *Wood Handbook* at each temperature over the range from 30°F to 210°F to the Hailwood and Horrobin model. Using equation (5.53) directly he calculated the values of K_1, K_2, and W, which would give the best fit to the sorption data for each temperature. This technique gave a better fit to the sorption data in each case than he obtained by solving for the intermediate constants A, B, and C by use of equation (5.54) and then calculating K_1, K_2, and W by means of equation (5.55). From the individual temperature values of K_1, K_2, and **W**, Simpson also calculated equations of these three variables as functions of temperature with the results

$$K_2 = 3.730 + 0.03642\,F - 0.0001547\,F^2 \tag{5.56a}$$

$$K_2 = 0.6740 + 0.001053\,F - 0.000001714\,F^2 \tag{5.56b}$$

$$W = 216.9 + 0.01961\,F + 0.005720\,F^2 \tag{5.56c}$$

where F is the temperature in degrees Fahrenheit.

In Table 5.2 are listed K_1, K_2, and **W** as calculated by Simpson (1971) at the individual temperatures from $30°F$ to $210°F$ at $20°F$ intervals. Also shown are the values calculated from equations (5.56 a, b, and c) at the same temperatures. It should be noted that the data used to calculate these equations included many intermediate temperatures to those shown in Table 5.2. The corresponding values of 1,800/**W**, equivalent to M_1, are also listed in Table 5.2. It is clear that **W** increases and M_1 decreases with increasing temperature, as anticipated.

Table 5.2. Hailwood and Horrobin Coefficients Obtained from *Wood Handbook* Data Over a Range of Temperatures by Simpson (1971)

Temp. (°F)	Individual Temperature Calculations				Equations (5.56a, b, c)			
	K_1	K_2	W	1,800/W*	K_1	K_2	W	1,800/W*
30	4.15	.677	205	8.78	4.68	.704	223	8.07
50	6.13	.753	252	7.14	5.16	.722	232	7.76
70	5.80	.746	252	7.14	5.52	.739	246	7.32
90	5.87	.754	266	6.77	5.75	.755	265	6.79
110	5.81	.772	289	6.23	5.86	.769	288	6.25
130	5.91	.782	315	5.71	5.85	.782	316	5.70
150	5.49	.790	345	5.22	5.71	.793	349	5.16
170	5.93	.806	393	4.58	5.45	.803	386	4.66
190	5.58	.814	438	4.11	5.07	.812	427	4.22
210	3.82	.821	462	3.90	4.56	.820	473	3.80

*Value of 1,800/**W** is equivalent to $M_1(\%)$ by equation (5.52).

As indicated by equations (5.50) and (5.51), the moisture M in the wood can be separated into its two components, M_h and M_s, at any value for H or M. From equation (5.51), for example, the water of hydration M_h is given by

$$M_h = \frac{1,800\,K_1 K_2 H}{W(100 + K_1 K_2 H)} \tag{5.57}$$

and the water of solution M_s is given by

$$M_s = \frac{1,800\,K_2 H}{W(100 - K_2 H)}. \tag{5.58}$$

Figure 5.6 shows how the two components M_h and M_s vary with relative humidity H. It also shows the total sorption isotherm M. The sums of M_h and M_s at any value of H add up to the total moisture content M at the same humidity. The water of hydration M_h curve is similar to the shape of curve expected for the "type 1" sorption (see Stamm 1964 or Hearle and Peters 1960) often referred to as "Langmuir adsorption." It is characteristic of "chemisorption" or the sorption of gases on surfaces where a monolayer is formed, as discussed in the opening pages of this chapter. The heat of sorption is high for this type of sorption.

The water of solution M_s curve has a shape similar to that designated as "type 3" sorption. It is characterized by a negligible heat of sorption because the attraction of the water molecules for the wood is approximately the same as that of the water molecules for each other.

The total curve of M against H is the "type 2" designation in which several layers of solvent (water, in this case) are taken up by the sorbate (wood, in this case). It is characteristic of systems which exhibit multilayered sorption, and the attractive energy between the water molecules and the wood is higher than that between the water molecules in the liquid state.

In their original paper, Hailwood and Horrobin (1946) calculated the heats of hydration and of solution and the corresponding entropy changes for wool, silk, and nylon by consideration of the variation of the equilibrium constants K_1 and K_2 with temperature. The standard thermodynamic formulas (Hearle and Peters 1960) were used as follows

$$\Delta H_1 = RT^2 \, \partial \ln K_1 / \partial T \tag{5.59}$$

$$\Delta H_2 = RT^2 \, \partial \ln K_2 / \partial T \tag{5.60}$$

where ΔH_1 is the heat of hydration of water from the dissolved to the hydrated state and ΔH_2 is the heat of solution of the dissolved water from the liquid state, both on a molar basis. Thus for wool, using sorption data at 25°C and 40°C, they calculated $\Delta H_1 = -5,416$ and $\Delta H_2 = -1,056$ calories per mole of water, the total heat being the sum of the two, or $-6,472$. The negative sign indicates that heat is released during sorption from the liquid state. The heat Q generated per gram of water is calculated by dividing each of the above by 18, or $Q_1 = 301$, $Q_2 = 59$, and $Q_1 + Q_2 = 360$ calories per gram of water. This is somewhat larger than the value of Q_L at zero moisture content obtained for wool, to which it should correspond, according to theory.

Taking the same temperature range, 25°C (77°F) and 40°C (104°F), and using equations (5.56a) and (5.56b) for K_1 and K_2 as determined by Simpson (1971), the corresponding values of K_1 are 5.617 at 77°F and 5.844 at 104°F, and of K_2 are 0.7449 at 77°F and 0.7660 at 104°F. Substitution of these values into equations (5.59) and (5.60) gives $\Delta H_1 = +489$ and $\Delta H_2 = +345$ calories per mole. The corresponding values of Q_1 and Q_2 are $489/18 = 27$ and $345/18 = 19$ calories per gram of water. The sum of these, $27 + 19 = 46$ calories per gram of water, would be expected to approximate the value of Q_L (see Chapter 4), the differential heat of sorption for dry wood. This should be in the order of 250 to 300 calories per gram of water. The disagreement between these values indicates that either the Hailwood and Horrobin theory is in error in this respect or that the sorption data is poor. There is further disagreement in that the theory predicts that K_1 and K_2 should increase continually with temperature, which they do not, as is clear from equations (5.56a) and (5.56b), and from Table 5.2. The value of $Q_1 + Q_2$ for wool given above seems to be more nearly in the range expected, and perhaps better data on the sorption isotherms for wood are required to evaluate the validity of the Hailwood and Horrobin model equation in terms of the thermodynamic coefficients.

Although the BET and Hailwood and Horrobin theories appear to provide different explanations for the sorption isotherm for wood, the curves obtained can be related to each other. For example, the Langmuir curve for $n = 1$ is analogous to the hydrated water moisture content M_h of the Hailwood and Horrobin theory. The difference between the Langmuir curve (Figure 5.5) and the total curve $M - M_1$ is analogous to the dissolved water moisture content M_s of the Hailwood and Horrobin theory. Likewise, the value of M_m, the maximum moisture content in the monolayer, may correspond to the value of M_h at $p/p_0 = 1.0$ in the Hailwood and Horrobin theory. For example, the data obtained by Simpson for M_m and M_h at $h = 1.0$ from the *Wood Handbook* data as functions of temperature are listed in Table 5.3. It is clear that nearly the same values are obtained throughout the temperature range.

Table 5.3. Comparison of Values of M_m from the BET Theory
with Those of M_h at $H = 100\%$ from the Hailwood and Horrobin Theory
Based on *Wood Handbook* Sorption Data,
According to Simpson (personal communication)

Temp. ($^\circ$F)	M_m	M_h	Temp. ($^\circ$F)	M_m	M_h
30	6.45	6.19	130	4.84	4.68
50	5.91	6.12	150	4.40	4.22
70	5.87	5.88	170	4.18	3.79
90	5.47	5.52	190	3.51	3.39
110	5.14	5.11	210	3.14	3.00

Deformation in Space Sorption Theory of Malmquist

The sorption theory of Malmquist (1958, 1959, and 1967) treats the sorption of water by wood and other hygroscopic materials in terms of space-dimensional factors within the cell wall. It also considers the cohesive properties of the cell wall which might limit the swelling associated with water sorption and therefore the extent of water sorption by the wood. Finally, it takes into account whether sorption occurs on surfaces with two translational degrees of freedom or along one-dimensional lines corresponding to one translational degree of freedom. The original theory also considers other possibilities which are not treated here. These include three-dimensional sorption volumes, with three translational degrees of freedom, and specific sorption sites with zero translational degrees of freedom.

This theory considers that there is a sorption space within the cell wall which sorbed water molecules can occupy. There is also the surrounding space or vapor space, defined as the space which the water-vapor molecules occupy.

When the cell wall is fully saturated with water, the specific volume of the sorbed water molecule is taken as v_f', which is then also equal to the volume of a sorbed-water space-cell in saturated wood. When the cell wall is not saturated with water, the specific volume of the sorbed water molecule is v', which is larger than the volume v_f'. An empty sorbed-water space-cell is an empty cell in the sorption space. There are always some empty sorbed-water space-cells when the cell wall is not saturated.

The number of empty sorbed-water space-cells per molecule of sorbed water is equal to $v'/v_f' - 1$. For example, if the wood is half-saturated, this expression reduces to unity since v' is twice v_f'. This would mean that there is one empty space-cell for each full one.

Similarly, for the vapor space, the specific volume per water-vapor molecule is taken as v_f'' when the space is saturated with water vapor. When the vapor space is not saturated, the specific volume of vapor space per water-vapor molecule is designated as v'', which is therefore always larger than v_f''. An empty vapor space-cell is an empty cell in vapor space and is found only in unsaturated wood. The number of empty vapor space-cells per molecule of water vapor is equal to $v''/v_f'' - 1$.

Consider that the vapor-volume space represents that volume occupied by y^3 cubical empty vapor space-cells per vapor molecule in the vapor space. Therefore, y^3 is the number of empty vapor space-cells per molecule of water vapor, or

$$y^3 = v''/v_f'' - 1. \qquad (5.61)$$

Returning now to the sorption space, the empty sorption space-cells may be concentrated either on sorption surfaces (two dimensions of translational freedom) or on sorption lines (one dimension of translational freedom). There are also the possibilities of sorption on specific sorption sites (zero degrees of translational freedom), and in sorption volumes (three dimensions of translational freedom), which are not treated here.

For the sorption surface, $ny^2 = v'/v_f' - 1$ where n is the number of layers of molecules which can be sorbed on the surface. For the sorption line, corresponding to a long molecule with sorption sites distributed along its length, $ny = v'/v_f' - 1$.

In general, for the sorption space, the number of empty space-cells per molecule of sorbed water can be written

$$ny^i = v'/v_f' - 1 \tag{5.62}$$

where i designates the number of degrees of translational freedom of the sorbed molecules and n is the maximum number of layers permitted.

Combining equations (5.61) and (5.62) and eliminating y results in

$$v'/v_f' = 1 + n\,(v''/v_f'' - 1)^{i/3} \tag{5.63}$$

But v'/v_f' is equal to c_f/c, so equation (5.63) becomes

$$c_f/c = 1 + n((v''/v_f'') - 1)^{i/3} \tag{5.64}$$

where c_f/c is the ratio of the concentration c of water in the cell wall to the concentration c_f at full saturation.

Cohesion Factor

Up to this point the theory has not considered the cohesive properties of the wood which might limit the sorption of water. This factor is accounted for by introducing a term designated by Malmquist as the "cohesion factor," defined as follows

$$k_c = \frac{[1 - (p_f/p_0)]}{[1 - (c/c_f)]} \tag{5.65}$$

where p_f is the apparent vapor pressure of the sorbed water in the cell wall when the wood is at the saturation concentration c_f. In general, the ratio p_f/p_0 is less than unity because of the cohesive forces in the cell wall, according to Malmquist (1959).

If the cohesive factor is zero, there is no interference between the layers of water forming on the various surfaces within the cell wall. If the cohesive factor k_c is unity, the surfaces are so close to each other that there is no room for water molecules to become sorbed.

Equation (5.64) can be written in terms of vapor pressures p and p_f in place of specific volumes v'' and v_f'' by use of the ideal gas law, thus reducing to

$$c_f/c = M_f/M = 1 + n\,[(p_f/p) - 1]^{i/3}. \tag{5.66}$$

Combining equations (5.65) and (5.66) and rearranging gives

$$M_f/M = c_f/c = 1 + n \left\{ \frac{p_0}{p} \left[1 - k_c \left(1 - \frac{c}{c_f} \right) \right] - 1 \right\}^{i/3} \tag{5.67}$$

which reduces for $k_c = 0$ to

$$M_f/M = c_f/c = 1 + n\left(\frac{1-h}{h}\right)^{i/3} \tag{5.68}$$

and for $k_c = 1$ to

$$M_f/M = c_f/c = 1 + n\left(\frac{c_f/c}{h} - 1\right)^{i/3} \tag{5.69}$$

where h is the relative vapor pressure p/p_0.

Figure 5.7 explains the meanings of the cohesive factor k_c. When the value of $k_c = 1$, the sorption space surfaces A and B are so close that water cannot penetrate to one surface or the other without forcing the wood to swell. When the value of $k_c = 0$, the sorption space surfaces are so far apart that even when all n layers of water coat the surfaces there is no restraint or swelling of the wood. When k_c is between 0 and 1, there is some swelling and some interference between the water layers sorbed on surfaces A and B.

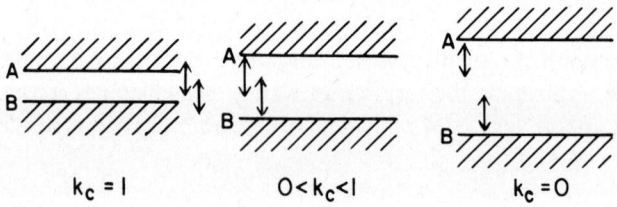

$$k_c = 1 \qquad\qquad 0 < k_c < 1 \qquad\qquad k_c = 0$$

Fig. 5.7. Schematic diagrams illustrating the meanings of the cohesion factor k_c (adapted from Malmquist 1959).

Figure 5.8, a graphical plot of equation (3.65), shows how p_f/p_0 varies with the ratio of M/M_f for various values of k_c. Note that when $k_c = 0$, there is no difference between p_f and p_0.

Figure 5.9 shows how the ratio M/M_f varies with relative vapor pressure h for $i = 2$ and for three different values of the cohesion factor k_c, according to Malmquist (1959).

The calculation of Simpson (personal communication) from the sorption data of the *Wood Handbook* show that the best fit to equation (5.68) results in a nearly constant value for i or 2.0 ± 0.1, over most of the temperature range 30–210°F. This indicates that the sorption surface is two-dimensional. The average number n of layers of water increased from approximately 2.5 at 70°F to about 3.7 at 210°F. The fiber-saturation point M_f decreased from 33 percent at 70°F to 25 percent at 210°F, consistent with expectations. It should be noted that equation (5.68) assumes that $K_c = 0$.

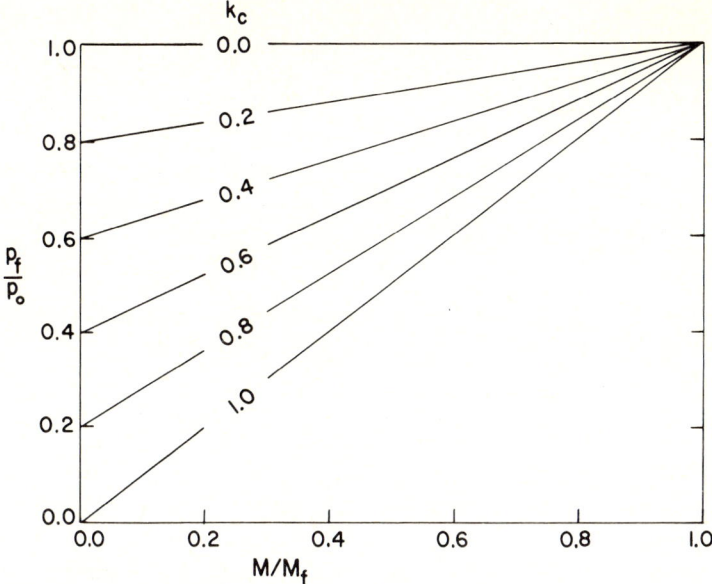

Fig. 5.8. Family of curves showing how p_f/p_0 varies with M/M_f for different values of the cohesion factor k_c (adapted from Malmquist 1959).

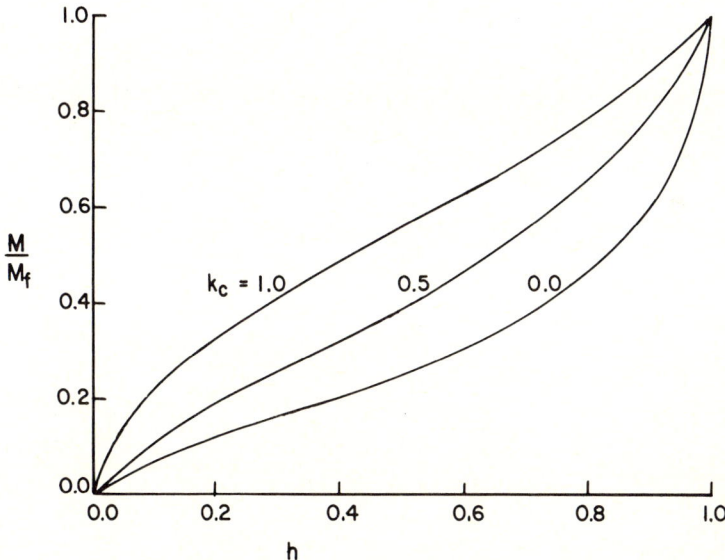

Fig. 5.9. Effect of the cohesion factor k_c on the sorption isotherms, plotted as M/M_f against h (adapted from Malmquist 1959).

Other Sorption Theories

As was indicated at the beginning of this chapter, a number of other theories have also been developed to explain the sorption of water by hygroscopic materials. Some of these theories have not been applied to wood, and it is of interest to determine if they are applicable. These theories will not be derived here, but the equations will be given in each case, as well as the general approach taken in their derivations.

Peirce Theory

One of the early sorption theories applied to textile materials is that of Peirce (1929). The derivation for this theory is given by King in Hearle and Peters (1960) and in Morton and Hearle (1962). This theory, in common with later ones, separates the sorbed water into two components, one (designated as alpha water) strongly attached to sorption sites in the material and the other (designated as beta water) attracted with smaller forces. The final isotherm can be written in terms of the fractional moisture content m_a representing the strongly attached alpha water analogous to the hydrated or monomolecular layer of water, and m_b representing the more loosely attached beta water, corresponding to the dissolved or multilayered water. Thus the relative vapor pressure h is given in terms of m_a and m_b as follows:

$$1 - h = (1 - Kwm_a) \exp(-Bwm_b) \qquad (5.70)$$

where k, $\mathbf{w}$, and B are constants. The constant $\mathbf{w}$ is equivalent to $\mathbf{W}/18$, where $\mathbf{W}$ has the same meaning as in the Hailwood and Horrobin theory; that is, the apparent molecular weight of the wood per mole of sorption sites.

According to the Peirce theory the two moisture components m_a and m_b are related to the total fractional wood moisture content m as follows

$$\mathbf{w}m_a = 1 - \exp(-\mathbf{w}m) \qquad (5.71a)$$

and

$$\mathbf{w}m_b = \mathbf{w}m - \mathbf{w}m_a = \mathbf{w}m - 1 + \exp(-\mathbf{w}m). \qquad (5.71b)$$

Simpson (personal communication) has fitted the *Wood Handbook* sorption data to the Peirce isotherm and has calculated the constants K, $\mathbf{w}$, and B as functions of Fahrenheit temperature F, with the results

$$K = 0.386 - 0.000606F \qquad (5.72a)$$

$$\mathbf{w} = 1.863 \exp(0.0152F) \qquad (5.72b)$$

$$B = 6.69 - 0.0579F + 0.000138F^2. \qquad (5.72c)$$

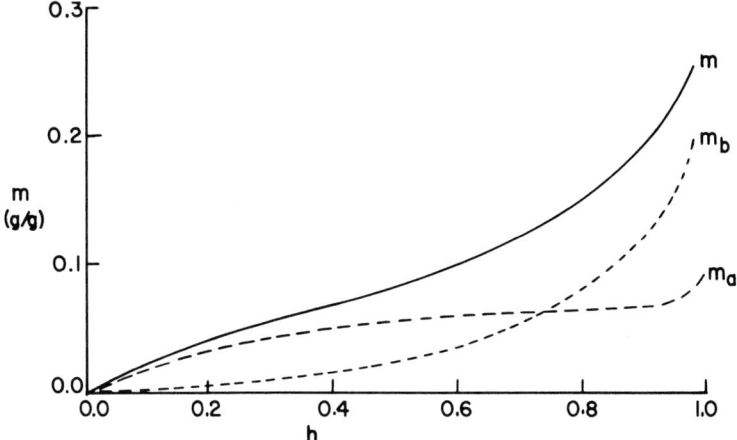

Fig. 5.10. Sorption isotherms calculated from the Peirce equation using *Wood Handbook* data at 110°F. Also shown are the alpha (m_a) and beta (m_b) components as function of h.

Figure 5.10 shows curves of m, m_a, and m_b as functions of h, calculated by Simpson at 110°F from the *Wood Handbook* data, using equations (5.70), (5.71a), and (5.71b). The original data points are also plotted for comparison with the curves calculated from the Peirce equation. The similarity of these curves to the Hailwood and Horrobin curves shown in Figure 5.6 should be noted.

According to equation (5.71a) the proportion of strongly bound moisture m_a decreases exponentially with moisture content m; that is, $1 - wm_a$ increases exponentially with increasing m. Peirce noted that this is similar to the relationship observed between the modulus of rigidity η of textile materials and moisture content. He therefore postulated that the rigidity modulus η was related to the amount of alpha water in the material. A similar exponential relationship applies to many of the mechanical properties of wood, represented by the following empirical equation

$$S = S_0 \exp(-Am) \qquad (5.73)$$

where A is a constant determined empirically from mechanical tests over the hygroscopic moisture range and S and S_0 are the strengths at the moisture content m and in the dry condition.

If we assume, as did Peirce, that the strength S is proportional to $1 - wm_a$, then

$$S = S_0 (1 - wm_a) = S_0 \exp(-wm) \qquad (5.74)$$

using equation (5.71). Therefore, comparing equations (5.73) and (5.74), it is clear that the coefficient A obtained from mechanical tests should be equivalent to the coefficient **w** obtained from the sorption isotherm. The value of **w** obtained by Simpson, equation (5.72b), at room temperature is approximately 7.5, using the *Wood Handbook* sorption data. It can be shown that this is approximately equal to the expected fractional change ΔS in S per unit fractional change Δm in moisture content. This is true because $\exp(-\mathbf{w}\Delta m)$ approaches $1 - \mathbf{w}\Delta m$ when $\mathbf{w}\Delta m$ is sufficiently small, as it is in this case, since $\Delta m = 0.01$. Therefore

$$(\Delta S/S)/\Delta m \approx -\mathbf{w}. \tag{5.75}$$

According to the *Wood Handbook*, the average value of $(\Delta S/S)/\Delta m$ (equivalent to $100(\Delta S/S)/\Delta M$, in $\%/\%$) observed in wood over most of the hygroscopic moisture range is a minus 2 percent for the parallel-to-grain Young's modulus per percent increase in wood moisture content M. In this case Young's modulus is less sensitive to moisture change than is predicted by equation (5.75). One may interpret this to mean that increasing alpha water content is therefore only effective in reducing wood stiffness by the ratio of the observed value $(\Delta S/S)/\Delta m$ of -2 to the predicted value of -7.5 at room temperature, or by $2/7.5 = 0.27$, or approximately 25 percent.

Other mechanical properties are more sensitive to moisture change. For example, the fiber stress at proportional limit and modulus of rupture in bending decrease by 4 and 5 percent respectively per percent moisture content. The relationship of mechanical strength to alpha moisture according to the Peirce theory is predicated on the assumption that the mechanical strength of wood is limited by the number of hydrogen bonds crosslinking the cellulose in the amorphous region of the cell wall. Sorption of alpha water presumably breaks hydrogen bonds on a one-to-one basis and therefore weakens the wood in proportion to the amount of alpha water present. If it takes four alpha water molecules to break each hydrogen bond, then one might expect the observed value of $(\Delta S/S)/\Delta m$ to be 2.

Further possible confirmation of the importance of alpha water in weakening wood is that increasing the temperature increases the absolute magnitude of **w**, therefore predicting that $(\Delta S/S)/\Delta m$ should increase with increasing temperature. This would appear to agree with the findings of Sulzberger (1953) as reported by Kollmann and Côté (1968) with respect to Young's modulus of elasticity for wood.

Enderby-King Theory

King in Hearle and Peters (1960) gives an equation for the sorption isotherm which can be written in the form

$$m = (18/\mathbf{W}) \left(\frac{Ak_1 p}{1 + k_1 p} + \frac{Bk_2 p}{1 - k_2 p} \right) \tag{5.76}$$

or letting $m_1 = 18/W$ from equation (5.52)

$$m = \frac{Ak_1pm_1}{1 + k_1p} + \frac{Bk_2pm_1}{1 - k_2p} \tag{5.77}$$

where W is the apparent molecular weight per mole of sorption sites (as in the Hailwood and Horrobin equation), m_1 is the fractional moisture content corresponding to that at which each of the sorption sites contains one molecule of water, A and B are constants proportional to the number of sorption sites containing only a monolayer of single water molecules and a multilayer of water molecules, respectively, k_1 and k_2 are equilibrium constants between the external vapor pressure $p(=p_0h)$ and the sorbed water in the monolayer and multilayers, respectively.

Equation (5.76) is similar to one derived by Enderby (1955), also based on the assumption of two types of sorption sites each of different binding energy (Figure 5.11). It is also apparent by comparing equation (5.76) with equation (5.50), the Hailwood and Horrobin equation, that they are also strikingly similar.

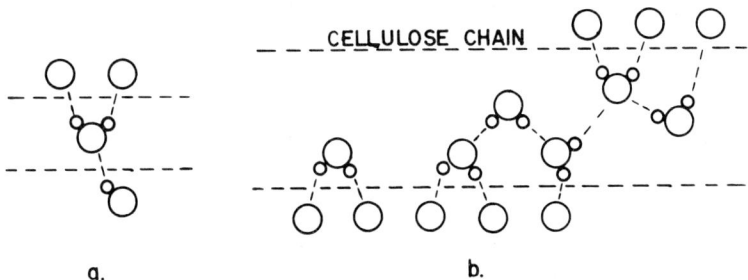

CELLULOSE CHAIN

a.   b.

Fig. 5.11. Schematic diagram showing the two types of sorption sites, primary (a.) and secondary (b.), and how the water is held in each case between the cellulosic chains (adapted from Enderby 1955).

Simpson (personal communication) has fitted the sorption data of the *Wood Handbook* to the Enderby-King equation and found the constants as functions of temperature from $F = 30°$ to $210°$ Fahrenheit

$$W = 225 - 0.0482\,F + 0.00543\,F^2 \tag{5.78a}$$

$$A = 1.17 + 0.00931\,F \tag{5.78b}$$

$$B = 0.921 - 0.00168\,F \tag{5.78c}$$

$$k_1 = 6.81(10^{-4})\exp(-0.0280\,F) \tag{5.78d}$$

$$k_2 = 5.05(10^{-4})\exp(-0.0383\,F). \tag{5.78e}$$

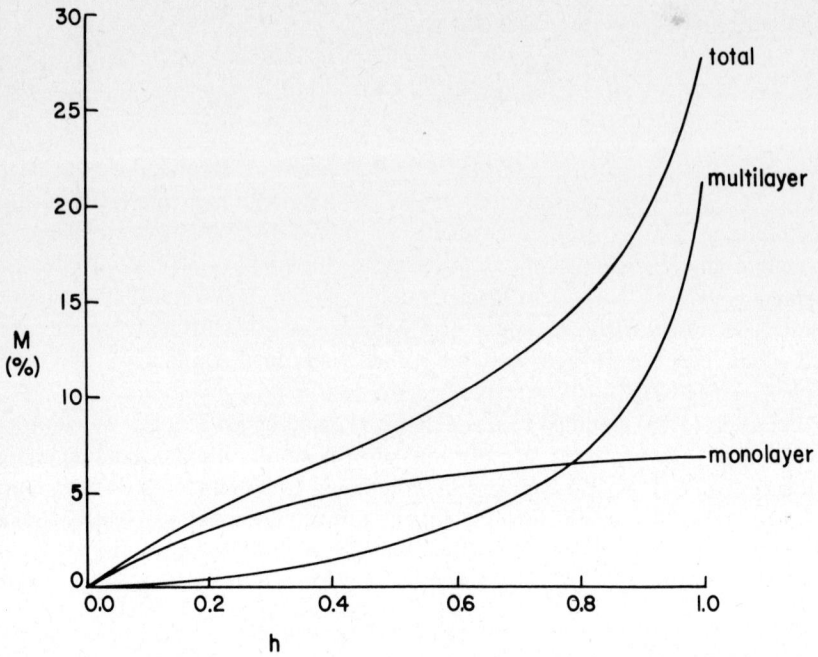

Fig. 5.12. Sorption isotherms calculated from the Enderby-King equation using *Wood Handbook* data at 110°F. Also shown are the curves for the monolayer and multilayer components.

The values of W obtained over the same temperature range by equation (5.78a) are nearly identical with those obtained by Simpson from the same sorption data using the Hailwood and Horrobin equation. Figure 5.12 shows the two terms in equation (5.77) plotted as functions of relative vapor pressure h, from data obtained at 110°F. The two lower curves represent the moisture content of the monolayer and of the multilayer, respectively, corresponding to the hydrated and dissolved water of the Hailwood and Horrobin model.

The constants k_1 and k_2 as stated above represent equilibrium constants between the water vapor and sorbed water in the monolayer and multilayer form, respectively. They are related to the equilibrium constants of equations (5.5a, b, and c) of the BET theory and to the energy of sorption U in the same way. The energy U_1 corresponding to the heat of vaporization per mole of water sorbed in the monolayer was found by Simpson to be 8,350 calories per mole, and U_2 the heat of vaporization for the multilayer was 7,410 calories per mole from the *Wood Handbook* data. One would expect these to be higher than these values. For example, the mean molar heat of vaporization U_0 of liquid water at the mean temperature of 120°F($\approx$50°C) of the sorption data is 10,224 calories per mole (18 × 568) (see Table 1.2 in Chapter 1). The energy U_2 of the multilayer should be close to this, while U_1 should be higher (see BET theory discussion), the difference between them being equivalent to $U_1 - U_0$ of the BET theory, equation (5.36), or 18 Q_L. For example, King in Hearle and Peters (1960) reports values of $U_1 = 14,400$ and $U_2 = 12,700$ calories per mole obtained from sorption isotherms for viscose rayon, a reconstituted cellulose.

The difference $(U_1 - U_0)/18$ gives a calculated value for Q_L of (14,400 − 12,700)/18 = 94.4 calories per gram of water for rayon and (8,350 − 7,410)/18 = 52.2 calories per gram of water for wood (using Simpson's results). This is in reasonable agreement with the value of 64.4 obtained for the same data using the BET theory, equation (5.38), on the same wood data.

Bradley Theory

Bradley (1936) developed a sorption theory which has been applied to water sorption in textile materials (Morton and Hearle 1962). It is based on the dipole attraction between successive layers of water molecules sorbed by the material. This attraction between successive layers may be visualized by reference to Figure 5.13 which shows several layers of oriented dipoles, water molecules in

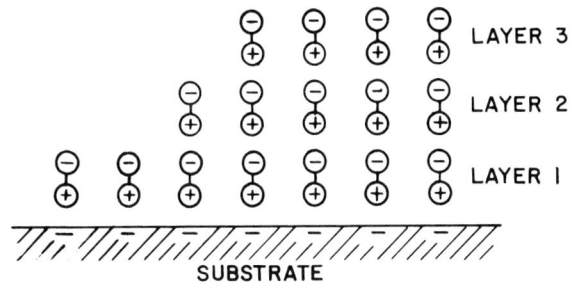

Fig. 5.13. Schematic diagram showing several layers of dipoles sorbed onto a negative substrate.

this case, attracted to a polar surface. The final equation derived by Bradley can be written as

$$\ln(1/h) = K_2 K_1^M + K_3 \qquad (5.79)$$

where M is the EMC in percent, h is the relative vapor pressure, and the constants K_1, K_2, and K_3 depend on the electric field strength of the sorption surface, the dipole characteristics of the sorbed water, and the temperature.

Simpson (personal communication) has also fitted the *Wood Handbook* sorption data to the Bradley equation and obtained the following empirical equations for the three constants as functions of temperature F in degrees Fahrenheit:

$$K_1 = 0.849 - 0.000236\,F \tag{5.80a}$$

$$K_2 = 3.64 + 0.00316\,F - 0.0000482\,F^2 \tag{5.80b}$$

$$K_3 = 0.00949 - 0.0000456\,F. \tag{5.80c}$$

It is clear from equations (5.80a, b, and c) that K_3 is small compared with the product $K_2 K_1{}^M$ except for large values of M. Therefore equation (5.79) can be approximated by

$$\ln(1/h) \approx K_2 K_1^M \tag{5.81}$$

or

$$\ln(\ln(1/h)) \approx \ln K_2 + M \ln K_1 \tag{5.82}$$

or in terms of log to the base 10

$$\log[\log(1/h)] \approx \log(K_2/2.303) + M \log K_1. \tag{5.83}$$

Figure 5.14 shows a plot of $\log[\log(1/h)]$ against M taken from the *Wood*

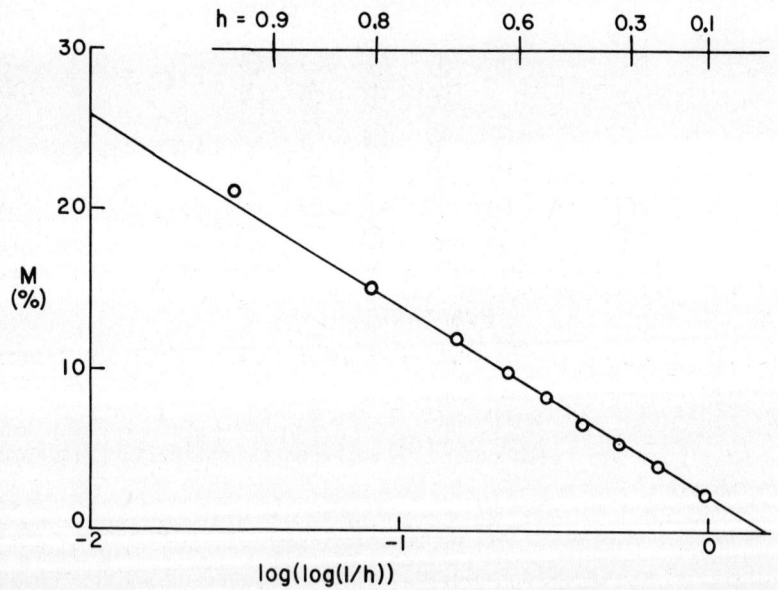

Fig. 5.14. Linear curve of moisture content M (%) against the logarithm of the logarithm of reciprocal relative vapor pressure h as predicted by the Bradley equation for wood at $100°F$ using *Wood Handbook* data (curve calculated by Simpson personal communication).

Handbook sorption data at 100°F using the values obtained by Simpson (personal communication).

A polarization theory of de Boer and Zwicker (1929) gives a sorption isotherm essentially identical to the approximate equation of Bradley, equation (5.81).

Kollmann Equation

Kollmann (1963) proposed that the sorption could be divided into three parts of which the total sorption m was the sum. Thus

$$m = m_1 + m_2 + m_3 \qquad (5.84)$$

where m_1 is the "true" sorption, given by

$$m_1 = A^h \qquad (5.85a)$$

m_2 is the submicroscopic capillary condensation, equal to

$$m_2 = C_1 \exp[-0.5(B_1 h - 1)^2] \qquad (5.85b)$$

and m_3 is the microscopic capillary condensation, or

$$m_3 = C_2 \exp[-0.5(B_2 h - 1)^2] \qquad (5.85c)$$

where A, B_1, B_2, C_1, and C_2 are constants and h is the relative vapor pressure. Figure 5.15 shows the total sorption isotherm and also the three components.

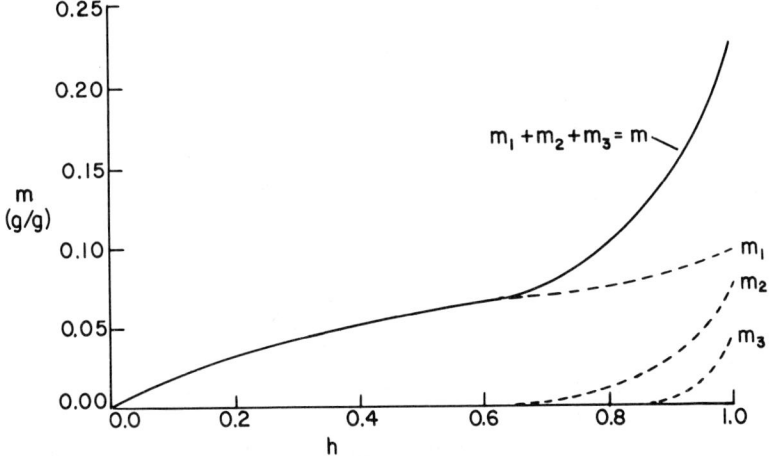

Fig. 5.15. Sorption isotherm and its three components as predicted by the Kollmann equation, where m_1 is the "true" sorption, m_2 the submicroscopic capillary condensation, and m_3 the microscopic capillary condensation (adapted from Kollmann 1963).

Additional Sorption Equations

Kadita (1960) proposed an equation of the form

$$m = \frac{Km_1 h}{1 + Kh} \left[1 + B\left(\frac{h - h^n}{1 - h}\right) \right] \qquad (5.86)$$

where K is an equilibrium constant related to the heat of sorption, m_1 is the moisture content equivalent to a complete monolayer, B is a constant related to multilayered sorption, and n is the number of layers in analogy with the BET equation. Kadita claims that his equation holds well for h greater than 0.05, and that the constants all have physical significance.

White and Eyring (1947) derived sorption equations for hygroscopic materials, based on statistical thermodynamics. In terms of moisture content m and relative vapor pressure h, these equations can be written

$$(m - m_a)(m_s - m_a) = Bm_a^2 \qquad (5.87a)$$

and

$$\ln[(m - m_a)/\dot{m}] + K_1 = \ln h \qquad (5.87b)$$

where m_a is the moisture content on localized sorption sites, m_s is the saturated moisture content of these sites, K_1 is the partial molar free energy change in the sorbed water due to the work of swelling, and B is the ratio of the partition function of the nonlocalized water $m - m_a$ to that of the localized water m_a.

Anderson and McCarthy (1963) used classical thermodynamic theory to obtain the following equation

$$h = \exp[(-18\,AB/RT)\exp(AM)] \qquad (5.88)$$

where A and B are constants and R and T are the gas constant and temperature, respectively. It can be shown that this reduces to equation (5.81), the approximate form of the Bradley equation, where $A = \ln K_1$, $18\,AB/RT = K_2$.

Doppert (1967) derived the following isotherm based on statistical thermodynamics

$$p = [k\theta/(1 - \theta)]\exp(2\theta\,u/RT) \qquad (5.89)$$

where p is the vapor pressure, θ is the ratio of the existing moisture content m to the saturated value m_f, k is a constant, $2u$ is the interaction energy between one sorbed water molecule and its nearest neighbors, R is the gas constant, and T is the Kelvin temperature.

Rowen and Simha (1949) applied thermodynamics to solution theory to obtain the following equation

$$\ln h = \ln V_1 + V_2 + \mu\,V_2^2 + \frac{KV_1'}{RT}\left(\frac{1}{V_2^{1/3}} - 1\right)\left(\frac{5}{3V_2^{1/3}} - 1\right) \qquad (5.90)$$

where V_1 and V_2 are the volume fractions of sorbed water and of wood, μ is a semi-empirical interaction parameter, V_1' is the partial molar volume of the sorbed water, K is a constant, and R and T are the gas constant and Kelvin temperature.

Stamm and Smith (1969) have developed a theory for the sorption and swelling of wood fibers on the basis of the cell walls being made up entirely of many concentric lamina consisting of chains of structural units with highly limited intra-laminar sorption and less limited inter-laminar sorption occurring so as to impose a minimum of stress within and between lamina. Sorption and swelling are confined to the outer hemicellulose and lignin zones of the structural units and the surface of their impermeable microcrystalline cores. Calculations give the best fit with sorption data and BET calculated sorption areas of contact when the structural units are assumed to have cross sections of 100 by 100 Angstrom units (cores 70.7 by 70.7 Angstrom units in cross section). These core dimensions are about twice those of the smallest microcrystalline unit observable with the electron microscope of chemically treated macerated fibers. These smaller crystalline units could very well be associated in unmodified wood. Intra-laminar sorption increases from an average of 0.6 to 2.8 water molecules thick per sorption site in going from a fiber specific gravity of 0.3 to 1.0. Intra-laminar sorption decreases from an average of 6.2 to 4.8 water molecules thick per sorption site over the same specific gravity range. These values are within the normally expected range. The theory also accounts for the tendency for the lumen to remain constant in size on swelling and shrinking without relying entirely on cross fibril outer and inner fibril wrappings (see Chapter 3).

Swelling Restraints and the Sorption Isotherm

The interaction of mechanical stress and sorption has already been discussed in Chapter 4 in connection with the hygroelastic or Barkas effect. It was pointed out that at constant relative vapor pressure a compressive stress reduces sorption and a tensile stress increases it. Barkas (1949) has calculated the sorption isotherms for spruce wood at room temperature for four different conditions of stress. These are shown in Figure 5.16.

The lowest isotherm in Figure 5.16 is for the case where sufficient stress is applied to the cell wall to maintain a constant cell-wall volume. This gives a fiber-saturation point m_f of only about 0.015 ($M_f = 1.5\%$). The next higher curve is for a constant volume of the gross wood; that is, all the swelling takes place into the cell cavities, giving $m_f \approx 0.13$. The third curve represents the natural sorption of the wood with $m_f \approx 0.24$. In this case the restraints to swelling are those caused by the wood structure itself. The fourth curve is calculated for "stress-free" sorption; that is, if the restraining sheath which reduces external swelling of the cell is removed. The extrapolated value of m_f is approximately 0.4.

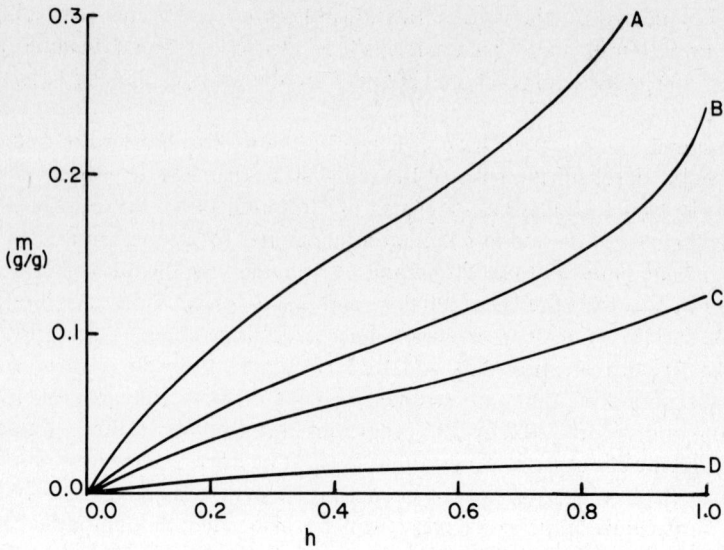

Fig. 5.16. Sorption isotherms proposed for wood by Barkas (1949) showing A. stress-free sorption, B. natural sorption, C. sorption under conditions of constant volume of the gross wood, and D. sorption under conditions of constant volume of the cell wall.

The stress-free curve includes the effect of internal stress within the cell wall exerted by the microfibrils in resisting swelling between and within them. As Barkas (1949) points out, if sufficient tension could be applied to the cell wall to overcome this internal restraint to swelling, the cell wall should go completely into solution as the relative vapor pressure h approaches unity, and m_f approaches infinity. Cassie (1945) has calculated curves for wool which are similar to those of Figure 5.16.

Literature Cited

Anderson, N. T., and Mc Carthy 1963. *Ind. Eng. Chem. Process Des. Develop.* 2:103 (original not seen, referred to by Venkateswaran 1970).

Andersson, R. 1952. Heat economy in timber drying. *Svenska Traforskningsinstitutet, Tratekniska Avdelingen No. 36B.* Stockholm.

Barber, N. F. 1968. A theoretical model of shrinking wood. *Holzforschung* 22(4):97–103.

Barber, N. F., and Meylan, B. A. 1964. The anisotropic shrinkage of wood. A theoretical model. *Holzforschung* 18(5):146–56.

Barkas, W. W. 1945. Swelling Stresses in gels. *Gt. Brit. Dept. Sci. Ind. Res. For. Prod. Res. Spec. Report No. 6.* London.

_____. 1949. The swelling of wood under stress. *Gt. Brit. Dept. Sci. Ind. Res. For. Prod. Res.* London.

Barrett, J. D., Schniewind, A. P., and Taylor, R. L. 1972. Theoretical shrinkage model for wood cell walls. *Wood Sci.* 4(3):178–92.

Beall, F. C. 1968. Thermal degradation analysis of wood and wood components. Ph.D. thesis. *SUNY College of Forestry at Syracuse University.* Syracuse, N.Y.

_____. 1969. Thermogravimetric analysis of wood lignin and hemicelluloses. *Wood and Fiber* 1(3):215–26.

Bello, E. D. 1968. Effect of transverse compressive stress on equilibrium moisture content of wood. *For. Prod. J.* 18(2):69–76.

Bosshard, H. H. 1956. Über die Anisotropie der Holzschwindung. *Holz als Roh-und Werkstoff* 14(8):285–95.

_____. 1963. Shrinkage and swelling anisotropy. *FAO Fifth Conf. on Wood Technology.* U.S. For. Prod. Lab. Madison, Wisc.

Bradley, R. S. 1936. *J. Chem. Soc.* 16:25 (original not seen, referred to by Morton and Hearle 1962).

Browne, F. L. 1957. Swelling of springwood and summerwood in softwood. *Forest Prod. J.* 8(11):416–24.

Brunauer, S., Emmett, P. H., and Teller, E. 1938. Adsorption of gases in multimolecular layers. *J. Am. Chem. Soc.* 60:309–19.

Busker, L. H. 1968. Microwave moisture measurement. *Instrum. and Control Systems* 41:89–92.

Cassie, A. B. D. 1945. *Trans. Faraday Soc.* 41:450 (original not seen, referred to by Morton and Hearle 1962).

Chalk, L., and Bigg, J. M. 1956. The distribution of moisture in the living stem in Sitka spruce and Douglas-fir. *Forestry* 26(1):6–21.

Christensen, G. N., and Hergt, H. F. A. 1969. Effect of previous history on kinetics of sorption by wood cell walls. *J. Polym. Sci. A1.* 7(8):2427–30.

Christensen, G. N., and Kelsey, K. E. 1959a. The rate of sorption of water vapor by wood. *Holz als Roh-und Werkstoff* 17(5):178–88.

_____. 1959b. The sorption of water vapor by the constituents of wood. *Holz als Roh-und Werkstoff* 17(5):189–204.

205

Clark, J., and Gibbs, R. D. 1957. Further investigations of seasonal changes in moisture content of certain Canadian forest trees. *Canad. J. Bot.* 35:219–53.

Cockrell, R. A. 1946. Influence of fibril angle on longitudinal shrinkage of ponderosa pine wood. *J. Forestry* 44:876–78.

_____. 1947. Explanation of longitudinal shrinkage of wood based on interconnected chain-molecule concept of cell-wall structure. *Trans. ASME* Nov.:931–35.

Cooper, D. N. E., and Ashpole, D. K. 1959. *J. Text. Inst.* 50 T-223 (original not seen, quoted by Venkateswaran 1970).

Crews, D. L. 1965. Structural variables and the differential transverse shrinkage of wood. Ph.D. thesis. SUNY College of Forestry at Syracuse University. Syracuse, N.Y.

De Boer, J. H., and Zwicker, U. 1929. *Z. Phys. Chem.* 133:407 (original not seen, referred to by Venkateswaran 1970).

Delgado, G. A. 1970. The effect of hygroscopicity on the heat of wetting of wood. Unpublished Senior Research Problem. SUNY College of Forestry at Syracuse University. Syracuse, N.Y.

De S. Wiejesinghe, L. C. A. 1959. The shrinkage of rays and fibres in wood. *Forestry* 32(1):31–38.

Doppert, H. L. 1967. *J. Polym. Sci., Part A-2.* 5:263 (original not seen, referred to by Venkateswaran 1970).

Duff, J. E. 1966. A probe for accurate determination of moisture content of wood products in use. *U.S. Dept. Agr. For. Prod. Lab. Res. note 0142.* Madison, Wisc.

Dunlap, F. 1912. The specific heat of wood. *U.S. Dept. Agr. For. Serv. Bul. No. 110.*

Enderby, J. A. 1955. *Trans. Faraday Soc.* 51:106.

Erickson, H. D. 1955. Tangential shrinkage of serial section within annual rings of Douglas-fir and western red cedar. *For. Prod. J.* 5(8):241–49.

Espenas, L. D. 1971. Shrinkage of Douglas-fir, western hemlock, and red alder as affected by drying conditions. *For. Prod. J.* 21(6):44–46.

Feist, W. C., and Tarkow, H. 1967. A new procedure for measuring fiber-saturation points. *For. Prod. J.* 17(10):65–68.

Forsaith, C. C. 1926. The technology of New York State timbers. *NYS College of Forestry at Syracuse University Tech. Publ. 18.* Syracuse, N.Y.

Frey-Wyssling, A. 1940a. The anisotropy of shrinkage of wood cross-sections. *Holz als Roh-und Werkstoff* 3:43–45.

_____. 1940b. The reasons for the anisotropic shrinkage of wood. *Holz als Roh-und Werkstoff* 3:239–53.

Goring, D. A. I. 1966. The structure of water in relation to the properties of wood constituents. *Pulp and Paper Magaz. Canada.* Nov. 1966 T:519–24.

Greenhill, W. L. 1936. The shrinkage of Australian timbers I. A new method of determining shrinkages and shrinkage figures for a number of Australian species. *Australian CSIRO Div. For. Prod. Tech. Paper No. 21.* Melbourne.

_____. 1944. The differential shrinkage of wood. *Trans. ASME* 66(2):152–54.

Hailwood, A. J., and Horrobin, S. 1946. Absorption of water by polymers: analysis in terms of a simple model. *Trans. Faraday Soc.* 42B:84–92.

Hann, R. A. 1969. Longitudinal shrinkage in seven species of wood. *U.S. Dept. Agr. For. Serv. Res. Note FPL-0203.*

Harada, H., and Wardrop, A. B. 1960. Cell wall structure of ray parenchyma cells of a softwood (*Cryptomeria japonica*). *J. Japan Wood Res. Soc.* 6(1):34–41.

Harris, J. M., and Meylan, B. A. 1965. The influence of microfibril angle on longitudinal and tangential shrinkage in *Pinus radiata*. *Holzforschung* 19(5):144–53.

Hearle, J. W. S., and Peters, R. H., eds. 1960. *Moisture in Textiles*. New York: Textile Book Publishers, Inc.

Hearmon, R. F. S., and Burcham, J. N. 1954. The dielectric properties of wood. *Dept. Sci. Ind. Res. For. Prod. Res. Spec. Rpt. No. 8*. London: Her Majesty's Stationery Office.

――――. 1955. Specific heat and heat of wetting of wood. *Nature* 176:978.

Henderson, L. G., and Choong, E. T. 1968. Variation in moisture content of standing sweetgum trees in Louisiana. *Louisiana State Univ. School of Forestry and Wildlife Mgt. Res. Release Note No. 81*. Sept. Baton Rouge, La.

Hergt, H. F. A., and Christensen, G. N. 1965. Variable retention of water by dry wood. *J. Appl. Polym. Sci.* 9:2345–61.

Hermans, P. H. 1946. *Contribution to the physics of cellulose fibers*. Elsevier.

Higgins, N. C. 1957. The equilibrium moisture content-relative humidity relationships of selected native and foreign woods. *For. Prod. J.* 7(10):372–77.

Hittmeier, M. E. 1967. Effect of structural direction and initial moisture content on swelling rate of wood. *Wood Sci. and Techn.* 1(1):109–21.

Hoadley, R. B. 1967. Weather, water and wood. *Univ. Mass. Coop. Ext. Serv. Pub. No. 15*. Amherst, Mass.

James, W. L. 1961. Calibration of electric moisture meters for jack and red pine, black spruce, paper birch, black ash, Eastern hemlock, and bigtooth aspen. *U.S. Dept. Agr. For. Serv. FPL Rpt. No. 2208*. Madison, Wisc.

――――. 1963. Electric moisture meters for wood. *U.S. Dept. Agr. For. Serv. Res. Note FPL-08. For. Prod. Lab*. Madison, Wisc.

――――. 1968. Effect of temperature on readings of electric moisture meters. *For. Prod. J.* 18(10):23–31.

James, W. L., and Hamill, D. W. 1965. Dielectric properties of Douglas-fir measured at microwave frequencies. *For. Prod. J.* 15(2):51–56.

Jimènez, A. 1967. The effect of stress on the vapor pressure of seven Venezuelan woods. Unpubl. Rpt. Dept. Wood Prod. Engin. SUNY College of Forestry at Syracuse University. Syracuse, N.Y.

Joule, J. P. 1859. On some thermodynamic properties of solids. *Phil. Trans. Royal Soc.* 149:91–131.

Kadita, S. 1960. Studies on the water sorption of wood. *Wood Res. Bul. Wood Res. Inst.* No. 23:1–61. Kyoto University. Kyoto.

Kauman, W. G. 1964. Zellkollaps im Holz-Erste Mitteilung:Einflussgrossen bei der Entstehung des Zellkollaps und seine Ruckbildung. *Holz als Roh-und Werkstoff* 22(5):183–96.

Kellogg, R. W., and Wangaard, F. F. 1969. Variation in the cell-wall density of wood. *Wood and Fiber* 1(3):180–204.

Kelsey, K. E. 1956. The shrinkage-intersection point—its significance and method of its determination. *For. Prod. J.* 6(10):411–16.

――――. 1957. The sorption of water vapor by wood. *Australian J. Appl. Sci.* 8(1):42–54.

――――. 1963. A critical review of the relationship between the shrinkage and structure of wood. *Australian CSIRO Div. For. Prod. Techn. Paper No. 28*. Melbourne.

Kelsey, K. E., and Clarke, L. N. 1956. The heat of sorption of water by wood. *Australian J. Appl. Sci.* 7:160–75.

Keylwerth, R. 1962. Untersuchen uber freie und behinderte Quellung von Holz-Erste Mitteilung:Freie Quellung. *Holz als Roh-und Werkstoff* 20(7):252–59.

――――. 1964. Untersuchungen uber freie und behinderte Quellung-Vierte Mitteilung:Untersuchungen uber den Quellungverlauf und die Feucktigkeitsabhangigkeit der Rohdichte von Holzern. *Holz als Roh-und Werkstoff* 22(7):255–58.

Koch, P. 1969. Specific heat of ovendry spruce pine wood and bark. *Wood Sci.* 1(4): 203–14.

――. 1971. Process for straightening and drying Southern pine 2 by 4's in 24 hours. *For. Prod. J.* 21(5):17–24.

Koehler, A. 1946. Longitudinal shrinkage of wood. *U.S. Dept. Agr. For. Prod. Lab. Rpt. No. R1093.* Madison, Wisc.

Kollmann, F. 1936. *Technologie des Holzes.* Berlin: Springer-Verlag.

――. 1959. Über die Sorption von Holz und ihre exacte Bestimmung. *Holz als Roh-und Werkstoff* 17:161–71.

――. 1962. Eine Gleichung der Sorptionisotherme. *Naturwissenschaften* 49(9):206–207.

Kollmann, F., and Côté, W. A. 1968. *Principles of Wood Science and Technology.* Vol. 1. Berlin: Springer-Verlag.

Kollmann, F., and Hockele, G. 1962. Kritischer Vergleich einiger Bestimmungsverfahren der Holzfeuchtigkeit. *Holz als Roh-und Werkstoff* 20(12):461–73.

Kubler, H. 1959. Studien uber Wachstumsspannungen des Holzes-Dritte Mitteilung:Langen-anderungen bei der Warmebehandlung frischen Holzes. *Holz als Roh-und Werkstoff* 17(3):77–86.

――. 1962. Shrinkage and swelling of wood by coldness. *Holz als Roh-und Werkstoff* 20(9):364–68.

――. 1970. A note on recovery of excessive shrinkage in wood. *Wood Sci.* 3(1):62–64.

Kuhlman, G. 1962. Investigation of the thermal properties of wood and particleboard in dependency on moisture content and temperature in the hygroscopic range. *Holz als Roh-und Werkstoff* 20(7):259–70.

Langmuir, I. 1918. *J. Amer. Chem. Soc.* 40:1361 (original not seen, referred to by King in Hearle and Peters 1960).

Lin, R. T. 1965. A study on electrical conduction in wood. *For. Prod. J.* 15(11):506–14.

Lindsay, F., and Chalk, L. 1954. The influence of rays on the shrinkage of wood. *Forestry* 27:16–24.

Lowery, D. P. 1971. Measurement of vapor pressure generated in wood during drying. *Wood Sci.* 3(4):218–19.

Lowery, D. P., and Kotok, E. S. 1967. Evaluation of a microwave wood moisture meter. *For. Prod. J.* 17(10):47–51.

Mac Lean, J. D. 1952. Effect of temperature on the dimensions of green wood. *AWPA Proceedings* 48:136–54.

Malmquist, L. 1958. Sorption as deformation in space. *Kyteknisk Tidskrift.* No. 4:1–11.

――. 1959. Sorption of water vapor by wood from the standpoint of a new sorption theory. *Holz als Roh-und Werkstoff* 17(5):171–78.

――. 1967. Untersuchungen zur empirisch-mathematischen Analyse der Sorption von Vasserdampf durch Holz. *Holz als Roh-und Werkstoff* 25(2):45–62.

Matsumoto, T. 1950. The anisotropic shrinkage of wood. Morioka Coll. Agr. and Forestry. Iwate Univ. *Bul. No. 26* (original not seen, cited by Crews 1965) (see also *Forestry Abstracts*, 1954. 15(1):92,654).

Mc Intosh, D. C. 1955. Shrinkage of red oak and beech. *For. Prod. J.* 5(5):355–59.

Mc Millen, J. M. 1950. Methods of determining the moisture content of wood. *U.S. Dept. Agr. For. Prod. Lab. Rpt. No. R1649.* Madison, Wisc.

――. 1958. Stresses in wood during drying. *U.S. Dept. Agr. For. Serv. For. Prod. Lab. Rpt. No. 1652.* Madison, Wisc.

Mc Millen, C. W. 1969. Specific heat of ovendry loblolly pine wood. *Wood Sci.* 2(2): 107–11.

Meylan, B. A. 1968. Cause of high longitudinal shrinkage in wood. *For. Prod. J.* 18(4): 75–78.

Morath, E. 1931. Swelling phenomenon of beech wood. *Kolloidchem. Beih.* 33:131 (cited by Kollmann 1936).

Morschauser, C. R., and Preston, S. B. 1954. The effect of ray tissue on transverse shrinkage of red oak. *Michigan Wood Technology. No. 1. April.* University of Michigan. Ann Arbor, Mich.

Morton, W. E., and Hearle, J. W. S. 1962. *Physical properties of textile fibers.* Manchester and London: The Textile Inst. Butterworths.

Myer, J. E., and Rees, L. W. 1926. Electrical resistance of wood with special reference to the fiber-saturation point. *NYS College of Forestry at Syracuse University. Tech. Bul. No. 19.* Syracuse, N.Y.

Nakato, K. 1958. On the cause of anisotropic shrinkage and swelling of wood. X. *J. Japan Wood Res. Soc.* 4(5):183–86 (cited by Crews 1965).

Nanassy, A. J. 1964. Electric polarization measurements on yellow birch. *Canad. J. Phys.* 42:1270–81.

Noack, D. 1964. Einfluss der Probenabmessungen auf die Bestimmung der Quellmasse von Holz. *Holz als Roh-und Werkstoff* 22(5):174–82.

Noack, D., and Kleuters, W. 1960. On the determination of moisture content of wood by means of radio-active isotopes (beta rays). *Holz als Roh-und Werkstoff* 8:304–308.

Panshin, A. J., and de Zeeuw, C. H. 1970. *Textbook of Wood Technology*, 3rd ed. New York: McGraw-Hill.

Parham, R. A. 1971. Crystallinity and ultrastructure of ammoniated wood. Part II. Ultrastructure. *Wood and Fiber* 3(1):22–34.

Parham, R. A., Davidson, R. W., and de Zeeuw, C. H. 1972. Radial-tangential shrinkage of ammonia-treated loblolly pine wood. *Wood Sci.* 4(3):129–36.

Paton, J. M., and Hearmon, R. F. S. 1957. Effect of exposure to gamma rays on the hygroscopicity of Sitka spruce wood. *Nature* 180:651.

Peck, E. C. 1950. Moisture content of wood in use. *U.S. Dept. Agr. For. Serv. For. Prod. Lab. Rpt. No. R1655.* Madison, Wisc.

_____. 1953. The sap or moisture in wood. *U.S. Dept. Agr. For. Serv. For. Prod. Lab. Rpt. No. 768.* Madison, Wisc.

Peirce, F. T. 1929. *J. Textile Inst.* 20:T133 (cited by Morton and Hearle 1962).

Pentoney, R. E. 1953. Mechanisms affecting tangential and radial shrinkage. *J. For. Prod. Res. Soc.* 3(2): 27–32.

Perkitny, T. 1963. Swelling pressure of wood. *Fifth F.A.O. Confer. on Wood Techn. U.S. For. Prod. Lab.* Madison, Wisc.

Petty, J. A. 1971. The determination of fractional void volume in conifer wood by microphotometry. *Holzforschung* 25(1):24–29.

Petty, J. A., and Preston, R. D. 1969. The removal of air from wood. *Holzforschung* 23(1):9–15.

Phillips, E. W. J., Adams, E. H., and Hearmon, R. F. S. 1962. The measurement of density variation within the growth rings in thin sections of wood using beta particles. *J. Inst. Wood Sci.* 20:64–66.

Pollisco, F. S. 1969. Physical properties of maple wood treated with ammonia vapor. Ph.D. Thesis. SUNY College of Forestry at Syracuse University. Syracuse, N.Y.

Pollisco, F. S., Skaar, C., and Davidson, R. W. 1971. Some physical properties of maple treated with ammonia vapor. *Wood Sci.* 4(2):65–70.

Prichananda, C. 1966. A study of some aspects of moisture sorption dynamics in wood. Ph.D. Thesis. SUNY College of Forestry at Syracuse University. Syracuse, N.Y.

Raczkowski, J. 1963. The swelling heat of wood. *Fifth FAO Conf. on Wood Techn. U.S. For. Prod. Lab.* Madison, Wisc.

Rasmussen, E. F. 1961. Dry kiln operator's manual. *U.S. Dept. Agr. For. Serv. Agr. Hdbk. No. 188. For. Prod. Lab.* Madison, Wisc.

Resch, H., and Ecklund, B. A. 1963. Moisture content determination for wood with highly volatile constituents. *For. Prod. J.* 13(11):481–82.

Ritter, G. J. 1939. Crystal arrangement and swelling properties of fibers and ray cells in basswood holocellulose. *Paper Trade J.* 108:33–37.

Ritter, G. J., and Mitchell, R. L. 1952. Fiber studies contributing to the differential shrinkage of cellulose. *Paper Indus.* 33(10):1189–93.

Rowen, J. W., and Simha, R. 1949. *J. Phys. Colloid Chem.* 53:921 (cited by Venkateswaran 1970).

Sadoh, T., and Christensen, G. N. 1967. Longitudinal shrinkage of wood. I. Longitudinal shrinkage of thin sections. *Wood Sci. and Techn.* 1(1):26–44.

Salamon, M. 1966. Effect of drying severity on properties of Western hemlock. *For. Prod. J.* 16(1):39–46.

———. 1971. Portable electric moisture meters for quality control. *Dept. Fisheries and Forestry. Canad. For. Serv. For. Prod. Lab. Rpt. No. VP-X-80.* Vancouver.

Schirp, M. 1968. Frostrisse an Baumstammen. *Forstarchiv* 39(7):149–54.

Schirp, M., and Kubler, H. 1968. Untersuchungen über die kältebedingten Längenänderungen kleiner Holzproben. *Holz als Roh-und Werkstoff* 26(9):335–41.

Schniewind, A. P., and Kersavage, P. C. 1962. Influence of rate of drying and rewetting on the dimensional changes of California black oak. *For Prod. J.* 12(1):29–33.

Siau, J. F. 1971. *Flow in Wood.* Syracuse, N.Y.: Syracuse University Press.

Simpson, W. T. 1969. Moisture changes induced in red oak by transverse stress. Ph.D. thesis. SUNY College of Forestry at Syracuse University. Syracuse, N.Y.

———. 1971a. Equilibrium moisture content prediction for wood. *For. Prod. J.* 21(5):48–49.

———. 1971b. Moisture changes induced in red oak by transverse stress. *Wood and Fiber* 3(1):13–21.

Simpson, W. T., and Skaar, C. 1968. Effect of restrained swelling on wood moisture content. *U.S. Dept. Agr. For. Serv. Res. Note FPL-0196.*

Sinnott, M. J. 1958. *The solid state for engineers.* New York: Wiley.

Skaar, C. 1948. The dielectric properties of wood at several radio frequencies. *Tech. Publ. No. 49* NYS College of Forestry at Syracuse University. Syracuse, N.Y.

———. 1964. Some factors involved in the electrical determination of moisture gradients in wood. *For. Prod. J.* 14(6):239–43.

Skaar, C., and Simpson, W. T. 1968. Thermodynamics of water sorption by wood. *For. Prod. J.* 18(7):49–58.

Smith D. M. 1954. Maximum moisture content method for determining specific gravity of small wood samples. *U.S. Dept. Agr. For. Prod. Lab. Rpt. No. 2014.* Madison, Wisc.

Spalt, H. A. 1957. The sorption of water vapor by domestic and tropical woods. *For. Prod. J.* 7(10):331–35.

———. 1958. The fundamentals of water vapor sorption by wood. *For. Prod. J.* 8(10):288–95.

Stamm, A. J. 1927. The electrical resistance of wood as a measure of its moisture content. *Indus. and Engin. Chem.* 19:1021–25.

———. 1930. An electrical conductivity method for determining the moisture content of wood. *Indus. and Engin. Chem. Anal. Ed.* 2:240–44.

———. 1935. The effect of changes in the equilibrium relative vapor pressure upon the capillary structure of wood. *U.S. Dept. Agr. For. Serv. For. Prod. Lab. Rpt. R1075.* Madison, Wisc.

———. 1955. Swelling of wood and fiberboards in liquid ammonia. *For. Prod. J.* 5(6):413–16.

———. 1960. Bound-water diffusion into wood in across-the-fiber directions. *For. Prod. J.* 10(10):524–28.

———. 1964. *Wood and Cellulose Science.* New York: Ronald Press.

———. 1967. History of two phases of wood science. *Wood Sci. and Techn.* 1(13):186–

———. 1971. Review of nine methods for determining the fiber saturation points of wood and wood products. *Wood Sci.* 4(2):114–28.

Stamm, A. J., and Arganbright, D. G. 1970. Surface tension of the sap of several species of wood. *Wood and Fiber* 2(1):65–66.

Stamm, A. J., Burr, H. K., and Kline, A. A. 1955. Heat stabilized wood (Staybwood). *J. Phys. Chem.* 39: 121–32.

Stamm, A. J., and Loughborough, W. K. 1935. Thermodynamics of the Swelling of Wood. *J. Phys. Chem.* 39: 121–32.

———. 1941. Variation in shrinking and swelling of wood. *Trans. ASME* Oct. 12–15. Louisville, Ky.

Stamm, A. J., and Seborg, R. M. 1935. Adsorption compression in cellulose and wood. I. Density measurements in benzene. *J. Phys. Chem.* 39:133–42.

Stamm, A. J., and Smith, W. E. 1969. Laminar sorption and swelling theory for wood and cellulose. *Wood Sci. and Tech.* 3(4):301–23.

Stevens, W. C. 1936. The transverse shrinkage of wood. *For. Prod. J.* 8(9):386–89.

Sulzberger, P. H. 1953. The effect of temperature on the strength of wood. *Aeron. Res. Cons. Comm. Rpt. ACA-46.* Melbourne (cited by Kollmann and Côté 1968).

Tarkow, H. 1960. Interaction of moisture and wood. *U.S. Dept. Agr. For. Serv. FPL Rpt. No. 2198.* Madison, Wisc.

Tarkow, H., and Turner, H. D. 1958. The swelling pressure of wood. *For. Prod. J.* 8(7): 193–97.

Tiemann, H. D. 1906. Effect of moisture on the strength and stiffness of wood. *U.S. Dept. Agr. For. Serv. Bul. No. 70* (cited by Tiemann 1944).

———. 1944. *Wood Technology.* 2nd ed. New York: Pitman.

Trapp, W., and Pungs, L. 1956. Einfluss von Temperatur und Feuchte auf das dielektrische Verhalten von Naturholz im grossen Frequenzbereich. *Holzforschung* (5):144–50.

Tsoumis, G. 1960. Untersuchungen über die Schwankungen des Feuchtigkeitsgehaltes von lufttrockenem Holz. *Holz als Roh-und-Werkstoff* 18(11):415–22.

Tsutsumi, J., and Watanabe, H. 1965. Studies on dielectric behavior of wood. I. Effect of frequency and temperature on ε′ and tan ∂. *J. Japan Wood Res. Soc.* 11(6):232–36.

Uyemura, T. 1960. Dielectrical properties of wood as an indicator of moisture. *Bul. Govt. For. Expt. Sta.* No. 119:95–172. Tokyo.

Venkateswaran, A. 1970. Sorption of aqueous and non-aqueous media by wood and cellulose. *Chem. Rev.* 70(6):619–37.

———. 1971. Application of dissociation hypothesis to electrical conduction in wood. *Wood Sci.* 3(3):183–92.

Vintila, E. 1939. Untersuchungen uber Raumgewicht und Schwindmass von Fruh-und Spatholz bei Nadelholzern. *Holz als Roh-und Werkstoff* 2(10):345–57.

Volbehr, B. 1896. Untersuchungen über die Quellung der Holzfaser. Dissertation Schmidt and Klaunig, Kiel (cited by Andersson 1952).

Vorreiter, L. 1963. Fiber saturation point and maximum moisture content of wood. *Holzforschung* 17(5):139–46.

Wagner, J. B. 1917. *Seasoning of wood.* New York: Van Nostrand (cited by Erickson 1955).

Walker, C. W. E. 1964. Microwave moisture measuring instrument. *I.S.A. Proc. 5th International Pulp-Paper Inst. Symp.* May 18–23.

Wangaard, F. F. 1969. Cell wall density of wood with particular reference to the Southern pines. *Wood Sci.* 1(4):222–26.

Wangaard, F. F., and Granados, L. A. 1967. The effect of extractives on water-vapor sorption by wood. *Wood Sci. and Techn.* 1(4):253–77.

Weatherwax, R. C., and Stamm, A. J. 1946. The coefficients of thermal expansion of wood and wood products. *U.S. Dept. Agr. For. Serv. For. Prod. Lab. Rpt. No. R1487.* Madison, Wisc.

Weatherwax, R. C., and Tarkow, H. 1968. Density of wood substance. Importance of penetration and adsorption compression of the displacement fluid. *For. Prod. J.* 18(7):44–46.

Weichert, L. 1963. Investigations on sorption and swelling of spruce, beech, and compressed beech wood between 20° and 100°C. *Holz als Roh-und Werkstoff* 21(8):290–300.

White, H. J., and Eyring, H. 1947. Adsorption of water by swelling high polymeric materials. *Textile Res. J.* 17(10):523–53.

Wilfong, J. G. 1966. Specific gravity of wood substance. *For. Prod. J.* 16(1):55–61.

Wood Handbook 1955. *U.S. Dept. Agr. For. Serv. Hdbk. No. 72. For. Prod. Lab.* Madison, Wisc.

Ylinen, A., and Jumppanen, P. 1967. Theory on the shrinkage of wood. *Wood Sci. and Techn.* 1(4):241–52.

Yokota, T., and Tarkow, H. 1962. Changes in dimension on heating green wood. *For. Prod. J.* 12(1):43–45.

Index